AI for Families

Ultimate Guide to Mastering AI

Shannon Kimberly Edwards

Contents

Preface

My favorite book when I was young was Madeleine L'Engle's *A Wrinkle in Time.* I was enthralled by the magic and mystery of space, science, and the idea of a smart, independent, young heroine bravely and confidently saving her family. In many ways, it reflects the vision I have for how my own kids might approach the future—bold, fearless, and full of imagination. If the youngest generations can stay enchanted by the magic and possibility of an unknown future while remaining committed to the fundamental truths of our humanity, then I think we'll all be okay.

I've spent much of my career telling stories about technology, sometimes as a writer and other times as a leader of consumer technology startups. In all cases, I have enthusiastically chased the newest and most exciting innovation, always wanting to be at the center of what's next.

For more than two decades, I worked for companies that used foundational artificial intelligence (AI) technologies in their offerings—such as product and movie recommendation platforms and online shopping search engines. And over the past ten years, I've started to advocate more regularly for AI, digital literacy, and related issues, such as data privacy, which I believe deserve far more commitment to protecting right now.

With the experiences I've had in and around the technology industry, one thing has been clear for some time: few people feel full ownership over the direction of their digital existence. And that is a problem. Because while technology's continued influence in our lives is inevitable, a passive approach to just letting it all happen is not.

We seem to have fallen sway to that inner voice that tells us we aren't qualified enough to speak up. Or that we don't live on a coast or in an influential city, or that we didn't attend the right school or study the appropriate subjects to have a say about AI. But it's this voice that's the issue, not the reality of an impending AI-driven world that needs to hear from all of us equally.

It's Always Been About AI

For as long as we've been on the Internet as consumers, and even decades before that, an AI-led existence has been the plan. Over the years, companies have collected more and more data, perfected their algorithms, and fed it all through increasingly powerful computers to get to where we are today. We are only now noticing because the ecosystem has become powerful enough to bring us consumer-facing generative AI tools such as ChatGPT.

It's always been the case that our data is essential to creating and improving these AI systems. More than essential, there is no AI as we know it today without large-scale human participation in an ecosystem that sucks in all of our rich, varied and diverse data.

But if my marketing experience has taught me anything, it's

that large numbers of consumers also have tremendous power when acting in concert. We have the power of the purse and strength in our numbers. It's not about withholding our information or creating conflict with technology companies; it's about acting more like critical and discerning stakeholders.

Can AI innovation make us smarter? Yes, I believe it can. Can these systems make us smart in the first place? I don't know. And that should be our starting point for considering and discussing AI's impact on children's lives. But to figure out the right approach, and especially related to kids' educational journey, will take many more voices than we are hearing from today. It's not just about speaking up regarding newer generative AI tools either, but the platforms that are now "amplified" and powered by AI technology.

We are the Bosses of These Machines

As we sit within the eye of the storm we have an incredible opportunity to reflect on what it means to be human and to "future proof" our families' lives. Because it's not just about being more technologically literate right now (although that's critical), but about strengthening and refining those skills that AI will never replicate or replace—like empathy, instinct, understanding, compromise, and love.

We also do ourselves a great disservice in meditating on all the ways that AI might take control of the future and even "destroy" humankind. It's far more likely that the future will be irreparably harmed by our lack of participation, ownership, and foresight into the many ways that AI impacts us along the way. If we sit back and let only a few decide, we may indeed reap what we have sown.

I hope in writing this book I can convince you that AI is not some hard-to-understand mystery but a simple-to-grasp technology, many decades in the making, that each of us is responsible for deploying, using and regulating. We can do great things with AI innovation, but we must consider every aspect of its potential impact. And not just on education and jobs, but our psyches, relationships, and deep understanding of who we are.

What Happens to Our Heroes?

We also have to reckon with the consequences of having already allowed ourselves to become mere data points to be organized and fed into AI models. As optimistic and excited as I am for the possibility that AI holds, I worry deeply about how we're choosing algorithmic order over renegade individuality. The messy, complicated, deeply human parts of our lives—our creativity, relationships, and capacity for wonder—don't always fit neatly into AI training data sets and we need to understand the consequences of trying to force it. If we're not careful, we risk sacrificing exactly those attributes that make us irreplaceable and different from machines.

For instance, as humans throughout time, we've loved and admired those innovators, creators, glass ceiling-breakers, and "heroes" who see the world differently, make great discoveries, and even save lives. We call these special people "one of a kind," "one in a million," "unequalled," or "singular." But now consider that in purely data terms they are really "outliers"—or more specifically data points that disrupt the statistical purity of an AI data training set where harmony (and therefore homogeneity) is the goal.

If we chose a world that operates solely on data and statistical perfection, what happens to creativity, heroism, "out-of-the-box" thinking, and radical innovation? When the data used in AI models only reflects the past, how do we encourage great thinkers to dream up the future? It's not that these are impossible questions, but issues that can only be addressed by the widest array of diverse voices. The implications for AI moving in the wrong direction are not just "the kids might get lazy" but a complete erosion of our very humanity.

What can we do? We start by recognizing that AI is neither Godlike nor even particularly "futuristic" (when you consider, again, that the data used is from our past), but a great tool of efficiency to make humans work smarter and better. We shouldn't attribute to it a "mind" or some magic ability. AI innovation comes from humans; it's built on top of our data, and it should serve our collective benefit. AI is a means, not an end. So let's not give up on our heroes and our humanity that easily.

How the Book is Organized

You'll find three parts to this book. The first focuses on what AI is, where you'll find AI in your day-to-day lives, and how it's already in schools. In Part Two, we'll cover some of the risks and tips for keeping families aware and safe. And then in Part Three, we'll celebrate the opportunities AI will present in the years ahead.

You might notice that I generally refer to readers of the book as "families" (rather than "parents"). It's because everything I'm sharing is geared to any adult who feels responsible for society's youngest humans. From grandparents, aunts, uncles and teachers to mentors, coaches, and community leaders—we need everyone on board.

The thing is, I'm just one person with enough experience to know what questions we should also be asking right now and an idea of where to look for answers. But I'm no different from each one of you reading this book right now. We all bring unique skills, rich perspectives, and new ideas regarding AI to discussions about the future. And that's the point I hope you'll take away from *AI for Families*. Because if it gives you the confidence to have informed, empowered conversations at work, home, or school, then I'll consider the book a great success.

We need to balance the power between those building AI systems and the rest of us. We are the data that these systems are built from, and we are the consumers of whatever gets produced. There is no AI without all of us. When we start to recognize this fact and our collective responsibility for protecting and guiding the next generation, we can finally start celebrating what's possible with AI instead of just bracing for impact.

Part One

AI Fundamentals: What's Real, What's Hype, What's Here

Chapter 1
What is Artificial Intelligence?

When my now-teenager was in middle school, she was showing her grandmother how smartphone voice commands work, and her grandmother asked, "How does it know what I'm saying?" My daughter paused, realizing that our entire family used Siri dozens of times each day without ever considering how it actually worked.

Even though much of my career has been spent at companies built on foundational AI technologies, such as machine learning for product recommendations and algorithms that suggest movies to watch, my family has never really thought of ourselves as "AI users." What we do instead is mindlessly let Netflix choose what we should watch on a Friday night, follow GPS directions on weekend adventures, and purchase items recommended to us on Instagram. And yet, all of these long-familiar platforms are powered by technologies that actually make up the foundation of AI.

For years, this casual relationship with AI worked fine for families like mine. We could enjoy the conveniences of new technology without needing to understand what's "under the hood." But now the technology that once worked quietly in the background wants to have conversations with our kids. It's like we spent years repeating the warning about not talking to strangers online only to invite the ultimate stranger to move into our schools and homes.

The problem is that AI has become a catch-all term for a broader set of technologies that allow computers to simulate human intelligence. Most of us haven't really given much thought to what AI might mean to our lives, and we don't realize that we've been using aspects of AI for years. In fact, according to a 2024 survey, 99% of Americans, like my family, had used at least one "AI-enabled" product in the past week, such as social media, entertainment streaming services, and GPS.[1] But when also asked if they "used AI" during this same period, 50% said they had not, and 14% were unsure.

This disconnect shows how much we need a shared vocabulary to describe what's actually happening in our world right now. We're not just dealing with smarter versions of familiar platforms anymore, but now chatbots that are fully integrated and tasks that are handled autonomously in the background. And the impact goes well beyond any one tool, as AI technologies can systematically alter our fundamental understanding of truth, disrupt how we learn, and even change the ways we relate to one another.

But there is good news here. And lots of it. AI is remarkably easy to explain, and its applications are logical. A technology that seeks to replicate human intelligence actually makes each one of us an expert. And the time is now to start using that expertise to weigh in on its development.

A Shared Understanding of AI

AI is essentially a sophisticated pattern recognition system that mirrors how humans naturally process information. If you think about how we recognize patterns in our daily lives, you can start to understand how AI systems work. For instance, we often understand our child's mood from a simple facial expression or can sense that it's probably going to rain by looking at the clouds in the sky.

We use patterns to make these types of assessments and decisions every day, taking in countless data points, from smells and visual cues to the behaviors of others and historical context. And we process all of this information automatically. AI does this too, but on a far larger scale and at a speed and sophistication that grows each day. But even as AI systems output content that feels familiar, they are only mimicking human intelligence.

The goal of this first chapter is to encourage a healthy, realistic understanding of AI by building a solid foundation for what the technology is and what it is not. By better understanding these systems, we are less likely to cede to them authority over us. We'll approach AI from a position of curiosity rather than fear. While legitimate concerns exist about AI's impact on education, privacy, and child development, understanding how these systems work can empower families to make informed decisions rather than reactive ones.

AI innovation also introduces big philosophical questions that we'll wrestle with for years to come. On the plus side, this does present an opportunity for families to discuss personal values, culture, and thoughts about the future—topics we might not have otherwise had reason to explore with our children.

What You Will Discover in This Chapter

We'll start this chapter by exploring the history of AI development and why understanding the past can help families separate fact from industry hype. You'll learn key terms that will help you engage confidently in AI discussions, and you'll develop an understanding of how AI should complement rather than replace human capabilities. We'll also delve into the issue of why AI systems are sometimes wrong and how it happens, along with a review of what it means for AI systems to introduce bias into their output.

I hope that by the end of this chapter, you'll have a clear, foundational understanding of AI technology and feel motivated to dig into the subject further. Remember, there is no "right" answer when it comes to the role AI can and should play in our lives. But by feeling more confident and willing to discuss and debate AI innovation with your family, friends, colleagues, and neighbors, you can help ensure that the path ahead is one rooted in pragmatism and optimism.

So let's get started by better understanding the path that got us to this place where AI seems to be rapidly shaping (and sometimes upending) our lives. As they say (well, specifically as Shakespeare wrote in *The Tempest*), "what's past is prologue," and this is particularly true as AI's past developmental milestones and trajectory can provide us with a solid map to the years ahead.

A Brief History of AI Development

The seeming speed of AI's arrival into our daily lives can feel disorienting and overwhelming. It's like we collectively blinked, and all of a sudden our kids are asking chatbots to help with their homework, and we don't have the experience or personal context to know if that's even okay.

We're not alone in feeling this way either. When ChatGPT launched in November 2022, it not only surprised the public but also an entire industry of seasoned technology professionals. *New York Times* tech columnist Kevin Roose wrote just two months after ChatGPT's launch that an "AI arms race" had begun.[2] And now, just three years later, it appears that was perhaps an understatement.

What might come as a surprise, though, is that AI's "sudden" ascent wasn't all that immediate—and the industry response was more about the competitive and commercialization opportunities of ChatGPT versus AI overall. And this is because innovation has been steadily building through decades of progress. It's like how we barely seem to notice our children growing and changing until they tower over us and are "suddenly" off to college. We tend not to notice these incremental changes in life until the impact feels immediate and seismic (like our kids leaving home!).

But this is a critical point to consider as you think about AI's impact on our lives today and future prospects: AI innovation and its many familiar milestones (like the first "chatbot" created decades ago) have been in motion for more than 60 years. I'd even go so far as to say: AI innovation has always been the endgame, and we were just too distracted to notice.

Naming this Fast-Moving Innovation

AI innovation traces back to the 1940s and 1950s, when mathematicians like Alan Turing began asking whether machines could "think." Turing's famous 1950 paper, "Computing Machinery and Intelligence," introduced what became known as the "Turing Test," an early way to determine whether a machine could show intelligent behavior equal to a human (and as judged by another human).[3]

The phrase "artificial intelligence" was actually coined in 1956 at a Dartmouth College conference where academics explored how machines could simulate human learning.[4] They needed a catchy name for this technology that would mimic human intelligence, and it seems to have stuck.

The short, clever nature of the phrase has had its downsides, though. While easy to understand and remember, much of the breadth and complexity of AI is obscured by it sitting so tidily under one single phrase. But naming the technology was important as AI innovation was already accelerating in its development. Even before the Dartmouth meeting, the "IBM-Georgetown experiment" had successfully demonstrated computer translation, with a few dozen sentences converted automatically from Russian to English in 1954.[5] And there would be so much more to come from that point onward.

Meet the Original Chatbot, ELIZA

In the mid-1960s, long before Siri or ChatGPT, a computer scientist named Joseph Weizenbaum at MIT created a "chatbot" called ELIZA (after the character in Pygmalion).[6] ELIZA worked by using simple rules to recognize patterns in a person's text communication and then respond with phrases that sounded like the program was talking to the user. Essentially, the program used a basic script that turned the user's statement into a question. For instance, if you were to say, "I'm feeling stressed about my job," ELIZA might reply, "Why do you think you're feeling stressed about your job?"

What made ELIZA so meaningful to the history of AI was the fact that despite this type of "mimicked" response, users felt a connection to the program. And while Weizenbaum had hoped to use ELIZA to show the limitations of computers, he instead illustrated how quickly humans could assign empathy and intelligence to a machine, despite the user intellectually knowing otherwise.

Considering the clear implications of ELIZA decades ago, it's quite remarkable to think that conversations around the future risk of humans anthropomorphizing machines did not happen more broadly and far sooner. It's not as if there was a lull in development either; every decade saw chatbots that were built upon this early work, and we should have more publicly considered the risks at every juncture.

How Tabloid Drama Brought Attention to AI

During these early decades of AI innovation, the industry was often plagued by "hype" and irrational enthusiasm that would regularly outpace the limitations of what could be achieved.[7] This fact would affect financial investment and public awareness of the work. Even notable advancements, such as the 1980s era "expert systems" that replicated the decision-making of human specialists in fields like medicine and finance, were too "narrow" in capability to get many excited.

What it took was human drama to capture our attention again. By the 1990s, this arrived in the form of juicy human-machine tabloid-like conflict. The setting was a chess rematch in 1997 between IBM's Deep Blue computer and chess champion Garry Kasparov.[8] While Kasparov had won their first match in 1996, the computer won after a marathon rematch that following year. The fallout was dramatic.

Computer scientist Mark Robert Anderson later wrote that the event was just as much about spectacle as it was about technological achievement.[9] From disbelief that Deep Blue could make such "human-like" decisions to Kasparov's emotional reaction to losing—the match was supposed to highlight techno-logical prowess but instead sparked confusion and concern in its outcome. It would also foreshadow the existential struggle that these systems would bring with them in the years that followed.

Our Worst Technology Fears Now on Screen

The history of AI isn't just about niche computer science events but also about how it shaped our imagination, art, and culture. Hollywood started delivering dystopian fantasies with AI-like systems as film characters starting in the 1960s. The first movie to really articulate an AI system this way was Stanley Kubrick's 2001: A Space Odyssey.[10] To have dreamed up the human-like computer HAL 9000 shows us today how clear AI's potential was from the start.

Throughout the 1980s, we were treated to dozens of movies about AI's deadly consequences or the murderous robots it would power. Think *WarGames*, *Blade Runner*, and *The Terminator*, to name just a few. But while we were dreaming up new ways for robots to go rogue, the real AI ecosystem quietly grew in the background. And by the time we became enamored by the World Wide Web, we had lost sight of the fact that AI was advancing by leaps and bounds just beneath the surface.

As Dotcoms Went Bust, AI Was Booming

You might be surprised to hear that as we turned our attention to "the Internet," AI innovation was rapidly advancing in the background. In fact, during the 2000s, there was so much attention paid to the flashy rise and fall of dotcom companies that we didn't notice how the surviving businesses were building out the foundations of AI.

In fact, when I began working at Netflix in 2003, the company's algorithms for analyzing customer movie-watching behavior were already incredibly effective. And Google's search system was growing rapidly (by the end of 2000, there were 60 million Google searches being conducted each day).[11] Online shopping was increasingly intuitive, and consumers began to seek its convenience; GPS systems became part of our commutes; and music playlists were now customized to our tastes. Companies were gathering the ingredients for AI, and we didn't even notice.

By the time the Internet industry found its footing again, AI innovation was well-positioned to just keep growing. The real transformation to come would be from three important factors converging over the next decade: a massive increase in computational power, an explosion of digital data from all of these new websites we were using (and many of them for "free"), and algorithmic breakthroughs (or the "math" behind AI).

AI Finally Fully Takes the Stage

The 2010s marked AI's entry into the mainstream with new consumer-facing applications. Smartphones offered constant data streams that made AI interactions (and subsequent data generation and collection) possible on a much larger scale. Voice assistants like Siri on iPhones (2011) and Amazon's Alexa (2014) brought AI into our homes through "natural language processing." Social media platforms used AI to curate feeds and target ads to users. Smartphone cameras began utilizing AI for automatic focus and photo enhancement. And so on.

AI had now shifted from a specialized technology to an invisible infrastructure that supported much of our online lives. We embraced it enthusiastically, but perhaps without fully understanding the long-term implications or consequences. But ready or not, the future would arrive in 2022 with the launch of ChatGPT.

A Quickly Shifting Generative AI Landscape

What started with ChatGPT's surprise debut has quickly become a crowded field of new offerings and household names. In just its first three years, we went from one chatbot that caught the world's attention to dozens of AI platforms competing for our conversations and engagement.

Today's landscape includes major players like Google's Gemini, Anthropic's Claude, and Microsoft's Copilot, each with their strengths and approaches. But it's not just stand-alone chatbots anymore either. Generative AI is also embedded into platforms we use each day. Instagram and Facebook now have AI assistants, Snapchat offers an AI friend feature, Grok is part of X (formerly Twitter), and even our search engines have morphed into conversational interfaces.

There are now also tools that go beyond text conversation. AI systems can generate images, edit videos, compose music, write code, and analyze documents. Some platforms focus on specific tasks like homework help or creative writing, while others promise to be a user's "digital assistant."

For families, this rapid expansion means the AI landscape our kids are navigating is moving so quickly that they are onto the next thing before we've even gotten started. But while the choice we now have is dizzying, the fundamental questions about how we want AI to fit into our family's life remain the same. And that's where we'll focus our attention in the chapters ahead.

Review of Key AI Terms

One of the greatest obstacles to truly understanding AI is the lack of a common "language" for us to discuss this technology—combined with a natural tendency by adults not to question what we don't understand. As adults, we often wait until we "get the gist" of something rather than more overtly asking, which is the opposite of what young kids do naturally.

In fact, Harvard child psychologist Paul Harris found that children ask an average of 40,000 questions between the ages two and five, and then by middle school, most have all but stopped. Warren Berger explores this phenomenon in his book, *A More Beautiful Question*, where he advocates for a rethink as to how we prioritize the art of asking questions.[12] He notes that those who have bucked this trend and continue to commit to rigorous inquiry have produced some of society's greatest breakthroughs. Without questions, there would be no innovation, and we do ourselves an even greater disservice in the AI age by not staying curious.

A good start is by looking at simple AI-related definitions. Because without a shared vocabulary and understanding, we won't get very far in advocating for our kids, family, friends, and communities. This doesn't mean memorizing complicated technical definitions or becoming an expert overnight. But it does mean developing enough of a common understanding to feel comfortable weighing in on school policy, helping our kids think about the tools they are using in school, and even navigating our own changing workplaces.

The urgency to become AI literate is growing, and societal rifts in knowledge acquisition are also already widening. In a recent survey conducted to gauge basic AI understanding, researchers found notable income and education disparities, with college-educated individuals earning over $100,000 exhibiting higher rates of AI literacy.[13] This divide makes building a shared AI vocabulary an even more urgent task.

So to get us started, below are some of the key definitions that represent the foundation of AI literacy. These will help us tackle the more detailed issues in the pages ahead. Many of these terms are probably already familiar to you. I've also organized definitions by importance and category to make them easier to digest.

Core Concepts

You can think of these terms as the very basics you'll need to get you through any conversation related to AI:

Artificial Intelligence (AI)

These are computer systems designed to perform tasks that typically require human intelligence, such as understanding language, recognizing patterns, or solving problems. Rather than following pre-programmed responses, AI systems can adapt and make decisions based on the information they process. The federal government also has an official definition that you may see referenced in policies.[14] It defines AI as "a machine-based system that can, for a given set of human-defined objectives, make predictions, recommendations, or decisions influencing real or virtual environments."

AI Model

An AI model is the "brain" that powers AI systems like ChatGPT. Models are trained for different tasks and have different strengths. The challenge here is balancing accurate output with necessary adjustments, since raw data often lacks context. Developers add guidelines and constraints, but these modifications can sometimes introduce new, unintended biases into the system.

ALGORITHM

Algorithms are the step-by-step instructions that tell computers how to process information and make decisions. It's like a detailed recipe that guides the computer through problem-solving. When your navigation system calculates the fastest route home, it's following algorithms that consider traffic, distance, and road conditions.

MACHINE LEARNING

Machine learning covers the ways in which an AI system improves its performance by learning from examples and data. It's similar to how you learn to predict your teen's mood patterns by considering your hundreds of interactions with them. The more data a system processes, the better it becomes at specific tasks. You've been experiencing machine learning for years through personalized recommendations on shopping sites, streaming services, and the content you see in social media feeds.

NATURAL LANGUAGE PROCESSING (NLP)

NLP is the technology that helps computers understand, interpret, and generate human language. It's what allows AI systems to read text, understand what you're saying when you speak to Siri, translate languages, or have a conversation with a chatbot. Think of it as teaching computers to speak "human"—turning our messy, complex language into something machines can work with and respond to naturally.

COMPUTER VISION

This is the technology that gives computers the ability to "see" and understand images and videos, much like human vision. It's what allows your phone to recognize faces in photos, helps self-driving cars identify stop signs, or lets you deposit a check by taking a picture of it. Essentially, it teaches computers to look at visual information and make sense of what they're seeing, rather than just storing images as meaningless data. More than a decade ago, I led the development of a computer vision-driven shopping app, which shows how long these types of advancements have been in development and use.[15]

FOUNDATIONS OF AI

This is the sum of core technologies and methods that have been quietly running AI systems for decades. These include (as mentioned above) machine learning (which helps computers recognize patterns in data), natural language processing (which helps computers understand text), and computer vision (which helps computers interpret images).

Think of it like the foundation of a house—you don't see it, but it's what everything else is built on. These foundational technologies have been powering the recommendation algorithms on Netflix, the spam filters in your email, the voice recognition in Siri, and the photo tagging on social media for years.

What's new isn't these foundations themselves, but how they've been combined and scaled up to create new conversational AI systems like ChatGPT that feel so dramatically different from what we've experienced before. The foundations were already there; we just finally built something visible and accessible on top of them.

Types of AI Systems

Now that we understand the basics, let's explore the specific types of AI systems your family likely encounters frequently:

GENERATIVE AI

Chatbots today are the product of "generative" AI. This is

AI that creates new content, including text, images, music, or code. When you ask ChatGPT a question, it "generates" a new response tailored to your specific request rather than simply retrieving stored information like a Google search.

Large Language Models (LLMs)

Often used interchangeably with "generative AI," an "LLM" is a type of AI specifically trained to understand and generate human language by processing massive amounts of text. ChatGPT, Gemini, and Claude are all LLMs. While we frequently use "AI," "generative AI," and "LLM" interchangeably when discussing chatbots, LLM specifically refers to those systems that focus exclusively on interpreting text and delivering a written response (versus a picture or other type of visual output).

Multi-Modal AI

As AI systems become more sophisticated, generative models can increasingly use more than text to deliver responses. So, AI systems that process multiple types of information simultaneously, including text, images, audio, and video, are "multi-modal" in nature. Many current AI platforms are both LLMs and multi-modal, meaning they excel at language while also understanding and producing other types of content.

How AI Works in Practice

Finally, let's look at the practical elements that make AI function in your daily life:

AI Prompt

A "prompt" is simply a question you ask of an AI system. This is how we've come to understand our work with AI systems like ChatGPT, Claude, or Gemini. Good prompts share the same characteristics as the best questions we ask in our daily lives: they're specific rather than vague, provide necessary context, and clearly state what kind of response you want.

TRAINING DATA

When we "teach" an AI system, data is used. And "training data" is this information. It includes text, images, videos, and other data that help AI learn patterns and make predictions. The quality and scope of training data directly influence the quality of AI responses. And as we'll explore in later chapters, understanding how training data is collected matters for our privacy too.

DATA PRIVACY

What we mean when talking about "data privacy" has widened in scope with AI. This now also means the control families have over what personal information AI systems collect, how that data is used, and who has access to it. When you or your children interact with generative AI, these systems often collect information about conversations, preferences, and usage patterns. Understanding data privacy means knowing what information you're sharing and how companies might use it.

AI AGENT

AI "agents" are systems designed to take actions on your behalf rather than simply providing information. Virtual assistants like Siri, Alexa, and Google Assistant are common examples. When you ask these systems to set a timer, play music, or add items to your shopping list, the AI performs actual tasks rather than just answering questions. It's the difference between asking a friend for restaurant recommendations versus asking them to actually make a reservation for you.

AGENTIC AI

A more advanced form of AI (which you can think of as the next step past AI agents) are systems that can plan, reason, and take actions to accomplish complex goals with minimal human oversight. While a basic AI agent might schedule one meeting when asked, agentic AI could potentially coordinate an entire project timeline, book meeting rooms, send invitations, and adjust schedules based on conflicts.

Making a Commitment to Lifelong Learning

This list is by no means complete, and newer AI concepts and capabilities will continue to emerge regularly. The key is building comfort with learning new terms as they become relevant rather than trying to master everything at once. And for us as adults to embrace the same type of critical inquiry that we once had as children. Because being curious and seeking answers is both the optimal way to understand AI and the best method to use it. And now, armed with this shared vocabulary, we can start to explore in more detail how AI can sometimes get things wrong.

Why and How AI is Sometimes Wrong

For a technology movement with so much at stake, it's no surprise that any missteps quickly capture our collective attention. For far too long, the tech industry has heralded its "move fast and break things" ethos. At one point, it was even Facebook's motto (until 2014). But when it comes to AI, this approach has meant embarrassing mistakes that have been, at best, awkward, and at worst, harmful.

There are many reasons that AI can make mistakes, and understanding the mechanics of how, why, and what that looks like can help us better consider the output and consequences.

If we consider AI errors to be teachable moments rather than technological failures, we are more likely to stay vigilant regardless of whether these AI systems get more accurate as time goes by. These moments will also help our kids understand that intelligence, whether artificial or human, requires questioning, checking, and continuous learning. So let's start by delving into why AI systems sometimes produce inaccurate answers.

Technical Shortcomings Leading to Inaccuracies

While technical shortcomings are often responsible for AI's missteps, it's important to remember that the technology is advancing swiftly and, as a result, many of these issues are quickly being addressed.

Limited Examples Available

Today's AI systems are only as good as the data widely available to power them, and that can present a huge challenge. It can also produce an imbalance in the picture presented. For instance, if an AI system were trained mostly on examples from one group of people, it might technically produce accurate answers for some but not others. This was already the case with early voice recognition systems. In 2016, sociolinguist Rachel Tatman noted that Google's speech recognition technology didn't work as well for women as it did for men.[16] The reason was simply a lack of female voice training data available to properly develop the product.

These limitations extend beyond voice recognition to the many technologies that families use each day. For instance, if an AI tutoring system was primarily trained on examples from one educational system or in a different cultural context, it might not prove universally useful. It's a simple-to-understand problem with significant implications for product quality and usefulness.

LACK OF APPROPRIATE CONTEXT

A movie streaming platform may recommend a horror movie for family movie night because it matches the patterns it's learned from a user (in this case, Dad is a horror buff). But if a family doesn't change user profiles between sessions, the system may not have the larger context to understand that, with an 8-year-old at home, a horror movie wouldn't be appropriate family viewing.

Context also includes the cultural reality and personal values that AI systems simply cannot understand. If trained on a limited set of data, it's harder for the system to factor in the bigger picture needed for appropriate responses. Pulling from what's available online has obvious limitations, and it also poses challenges as developers try to adjust their models without introducing additional bias.

Another problem with trying to improve context is that by providing additional information, users often give up some of their privacy in return. So we have a double-edged sword when it comes to context and personalization that we'll dig into in more detail late in the book.

PATTERN-BASED FABRICATIONS

AI systems learn to generate content by identifying patterns in their training data but can struggle to distinguish between accurate and inaccurate information within those patterns. When an AI system encounters a question about a controversial topic, it could, in theory, weave together elements from conspiracy theories, legitimate research, and fictional narratives to create a response that sounds plausible but is factually incorrect.

While AI companies work to filter out false and misleading information from their training data, this process is imperfect. We're still in the early days of generative AI, and these systems can confidently present fabricated "facts" with the same tone of authority as with accurate information, making it difficult for users to distinguish between reliable and unreliable responses.

Search Data Limitations

Beyond pattern-matching problems, AI systems also face timing issues. AI can't know about events that happened after it was trained. It might provide you with information about a restaurant that closed last month or offer outdated health guidance. And while newer versions of LLMs such as ChatGPT, Claude, and Gemini can increasingly access real-time information through web searches, they still have knowledge cutoffs and may not have the most recent data available to process.

On the other side of search, AI responses are now being integrated directly into search engines such as Google (which, to confuse you even more, is a different experience than if you went to Google's AI platform, Gemini, directly). When we do a Google search, we often have a certain way that we seek information and an expectation around the response. For instance, doing a quick search and expecting to be able to scroll through results. But now you have the equivalent of someone "butting in" and perhaps not understanding what you wanted or even providing inaccurate information.

The blending of AI into familiar tools can also create a false sense of reliability, making it even more important to maintain the same critical thinking we'd use with any AI-generated content.

Types of Inaccuracies in AI Output

As AI systems improve, inaccurate content will increasingly be deliberate rather than accidental. So it will be more important than ever to stay skeptical and trust your instincts when something feels off.

AI System "Hallucinations"

A lack of training data can mean that AI chatbots "fill in" any gaps in their knowledge with made-up information. When AI platforms generate plausible-sounding but incorrect information in this context, it's called a "hallucination." This gap-filling exercise has received a lot of attention in the news, as it's one of the most obvious ways that AI systems misfire.

Put simply, when an AI system can't fully answer a prompt, it won't just say, "I don't know." Instead, the AI chatbot creates convincing-sounding but entirely fabricated information to ensure you receive a response. Because these responses can be so convincing, it's critical to ask for evidence to support any answer or claim.

Synthetic Media and "Slop"

Beyond creating false text information, AI systems can also generate realistic-looking fake images, videos, and audio content, collectively known as "synthetic media." Using the same pattern recognition approach we discussed, AI can create photos of people who don't exist, generate videos of real people saying things they never said, or produce audio recordings that mimic someone else's voice.

While this technology has legitimate uses in entertainment and education, it also means that children might encounter AI-generated content that looks real but is entirely fabricated. We'll talk in later chapters about how this is also the basis for deepfake creation and other seemingly real but fake (and harmful) content.

The term "slop" refers specifically to weird, distorted, or inappropriate imagery generated by AI systems. Slop can result from AI models that haven't been properly trained or filtered, or this type of content can be intentionally created. You've probably already seen examples of intentional slop, such as bizarre hybrid creatures and unrealistic depictions of everyday objects.

Deliberate Cyber Attacks

While many AI errors result from technological limitations, families also need to understand that AI can be deliberately manipulated by criminals.

Bad actors can create AI chatbots specifically targeting children, who may be more trusting of authoritative-sounding responses. These malicious systems might pose as educational tools while encouraging dangerous behaviors or pretending to be friends or popular online personalities to trick children into providing personal information. Cybercriminals can also poison training data with false or dangerous information, as computer scientists from the National Institute of Standards and Technology warned in a 2024 report.[17]

Be cautious of AI systems that ask for excessive personal information, encourage secrecy, refuse to identify their creators, or appear through suspicious links. Whether AI errors stem from technical limitations or malicious intent, the solution remains the same: developing strong critical thinking skills and maintaining healthy skepticism.

Benefits of Skepticism and Curiosity in Using AI

What makes false information coming from AI systems particularly challenging is the unprecedented speed and scale at which it can be created and distributed. Whether we're talking about information generated inadvertently by AI platforms or intentionally by bad actors, it's often being created and shared faster than fact-checkers can identify and debunk it.

But if there is good news here, it's that with a more in-depth understanding of the technical limitations of AI systems, we have a head start in protecting ourselves and our families. We also have more reason to apply those critical thinking skills. Teaching children to approach AI responses with curiosity and confidence rather than blind acceptance and fear transforms potential problems into learning opportunities. When children learn to ask "How do you know this?" and "What might this be missing?" or ensure they receive citations and back-up claims, they are practicing skills that will serve them well beyond their interactions with AI technology.

These same skills are also what's needed as we navigate the harder-to-identify issue of understanding how AI systems can reflect and amplify bias. An issue more subtle than flat-out misinformation, AI-generated bias can have serious repercussions for the technology's fair and equitable use, particularly in schools.

How AI Systems Can Introduce Bias

Even AI systems designed to be objective and educationally robust can inadvertently reflect and reinforce societal biases because outdated content can be used to train an AI model. But unlike the more obvious mistakes or intentionally inserted inaccuracies that we just reviewed, biases that sneak into an AI model and then into its chatbot output may be nuanced and tougher to root out.

It's not a new problem, of course. Historical text has always reflected the biases and beliefs of their time, but in the past, if this faulty information were in a school textbook, for instance, the school would no longer use that book. But today, if outdated, unfair, or incorrect information is delivered via a chatbot with its conversational, authoritative tone, it can be difficult to question. It can also be so subtle and distorted as to make it unnoticeable until the bias has caused harm.

But on the plus side, this challenge represents an opportunity for us to talk to our kids about bias and illustrate why we should always question and seek to verify any AI-generated output.

Scarcity and Human Error

It can be unsettling to think that the AI systems we've invited into our children's schools are essentially trained on the same messy internet content we have them avoid for schoolwork purposes. In fact, researchers suggest that only about 40% (and potentially as little as 10%) of web-based data, after adjusting for any duplicate content, is even of "acceptable" quality to train AI systems.[18] So you can understand how easy it might be for bias to creep in when developers are already limited in the information they have available to work with.

It's also important to understand that bias can be found even in reputable sources of information—like data research studies or, again, historical text. We aren't talking just about "garbage in garbage out." You can think of it like if your child learned to cook using just one cookbook, and even then, the cookbook only had three solid recipes to select from, of course it would color their opinion of what was "good" cooking. And that's the type of bias that is particularly difficult to control.

While engineers work to account for bias in their algorithms, it's a task fraught with complications too, especially as individual engineers and organizations naturally bring their own inherent biases to the work. The biases can be subtle and difficult to trace. They can also create a compounding effect where small prejudices in training data and well-intentioned efforts to correct these biases can create equally skewed AI responses.

Types of Potential AI Bias

AI bias can show up in different ways, and understanding these patterns helps you spot problems before they become embedded in the systems your family uses.

HISTORICAL, CULTURAL, AND GENDER BIAS

One of the trickier aspects of AI bias for families is how it can quietly reinforce outdated assumptions that we've worked hard to move beyond. AI training data often reflects the historical period and cultural context in which it was created, perpetuating outdated (and incorrect) ideas about gender roles, race, religious practices, and social structures.

Sometimes a simple nuance to the data can end up misrepresented and amplified, creating large-scale wrong assumptions. For instance, researchers analyzing text used to train AI systems discovered that because there have been objectively fewer female leaders throughout history, those who have been considered "failures" or "bad" leaders are regularly named specifically and frequently in historical text.[19] While comparatively the number of men with failures under their belt represents a far higher number, researchers found that individual failed male leaders are mentioned less frequently. The consequence is that the AI systems in the study ended up interpreting the data to suggest that women might be less suited to lead.

The answer isn't to rewrite history, of course, but we also don't want past inequality to shape future possibilities. It's just one of the many ethical issues that needs to be addressed in the years ahead. But for today, we need to teach our kids that historical limitations don't define current or future potential and that AI responses must be viewed critically through this lens.

RACIAL BIAS

Racial bias in AI systems is one of the most pervasive and troublesome issues families face with this technology, and it shows up in ways that can be both subtle and deeply harmful.

AI systems trained on historical text or even recent data can perpetuate stereotypes via their suggestions and recommendations. Researchers have found that when AI systems recognize race-related dialect, for instance, they can generate output that reinforces negative stereotypes about academic ability, job performance, or behavioral characteristics.[20]

A lack of diverse visual training data, representing differences within a group, can also result in output that reinforces stereotypes. Researchers found that a search for a "Middle Eastern man" regularly produced individuals with beards and in traditional attire.[21] The same study found that when attributes were added to an image search via a generative AI tool, the output was even more biased and damaging. Prompts to deliver an image of "pilot," "accountant," or "journalist," for instance, resulted in images of a White person (and primarily a man) more than 75% of the time. While if the researchers asked for a "criminal," someone who was "poor," or a "musician," more than 50% of the images delivered were of someone who was Black (and for the latter search, fewer than 25% in each case were White and zero responses were of a woman).

What makes this particularly concerning is that children often perceive AI responses as more objective or authoritative than human opinions. Racial bias in AI output not only perpetuates harmful stereotypes, but it can also limit the opportunities of children who don't "see" themselves in the output.

GEOGRAPHIC AND LANGUAGE LIMITATIONS

For families from diverse cultural backgrounds, AI bias can show up in ways that feel quite personal—from inaccurately representing a family's culture or values to not providing results in the language they speak. Again, AI systems tend to reflect the cultural biases and geographic distribution of their training data. And since much of generative AI development happens primarily in English and in Western contexts, the technology can be less accessible and accurate for many families around the world.

It's not a problem simply relegated to AI output either. With fewer than 5% of the roughly 7,000 languages spoken around the world today having any meaningful online presence, not only does it exclude users, but over time this absence of language diversity contributes to what researchers call "digital language death."[22] It's incredible to think that living more regularly online can lead to entire languages disappearing —and the implications for the survival and thriving of different cultures in the future.

ECONOMIC AND RESOURCE ASSUMPTIONS

When looking for suggestions that are more local or related to travel, the data available to AI systems may create a mismatch with a user's expectations too. For instance, a lot of online content about travel, restaurants, and local resources tends to be more "aspirational" than practical in nature. Disproportionate marketing spend on specific locations to increase their web presence can also result in location content showing up with more frequency or prominence in an AI system's training data.

The result can be out-of-touch output. For instance, if a child asks an AI tutor for help with a science project, the system might cheerfully recommend visiting a science museum three hours away or purchasing specialized materials that cost more than your weekly grocery budget. Rural families might receive suggestions for services that simply don't exist in their area, while families with limited resources might get recommendations that are ridiculously beyond their financial means.

These assumptions can be particularly challenging when

you consider that producing better, more localized recommendations often comes at the expense of our privacy. For many, providing additional socioeconomic or geographic detail is not something they are willing to do and for good reason—something we'll explore in more detail in Part Two.

COMMERCIAL PERSONALIZATION BIAS

One of the ways AI can unduly influence families is through content personalization that appears helpful but ultimately creates echo chambers that can end up influencing behavior. It's an AI-driven version of the age-old "but everyone is doing it" saying. Researchers have found this to be particularly problematic when it comes to online shopping.[23] AI-amplified shopping ecosystems can now more significantly impact consumer decision-making and contribute more readily to impulsive spending.

Many families have experienced this firsthand, and particularly with their kids. For example, with the sudden (and worrisome) tween obsession over expensive, typically anti-aging skincare products.[24] The trend has become so pronounced that California legislators have actually sought to ban the sale of certain anti-aging cosmetic products to children under 13.[25] When you layer in AI to this already fraught shopping ecosystem, the idea that "everyone" is doing something can become even harder for families to fight (and be harder on our wallets too).

Teaching Kids to Recognize AI's Biases

These are only a few examples of the ways that AI systems can introduce bias—and we should always actively consider the possibility of bias in AI output. How can families address this problem? Rather than oversharing with kids the complexities of AI bias, a good start is in helping them sharpen their analytical instincts. Simple questions like "Who might disagree with this statement?" or "What gaps in history might account for this answer?" can help children better understand the issue at hand.

Conversations about what's missing from AI's cultural representation can also make for excellent family discussions. Talking about bias helps us better appreciate what makes human intelligence different from artificial intelligence. These are some of my favorite discussions to have with my own children, as each of us is unique and critical to the building of AI innovation now and in the future. It's further reinforcement that our full potential should only be enhanced by machines, not diminished by them.

Artificial vs. Human Intelligence

The human-like conversational tone of generative AI systems is both the technology's greatest strength and, as we've discussed already, what makes the technology unsettling and even outright problematic. When ChatGPT, Gemini, or Claude "talks" to us and responds in a way that "sounds" human, the output inherits the weight and authority that comes with being "like" us. It can also force into the daylight big, uncomfortable questions about what it means to be human at all.

But wrestling with this new reality also gives us a chance to have rich conversations with our kids that we might not otherwise have had a reason to launch into. When children notice that AI "talks like a person," we can explore what human intelligence actually means and what makes it so special. For instance, our capacity to dream, to care deeply about others, to find meaning in our individual experiences, and to create something entirely new from our imagination.

We want our children to understand early on that they remain in control and that AI will not replace their individual intellect or autonomy. It's a powerful tool, but still just a tool. Unlike humans, AI systems don't have genuine understanding or the creativity that comes from our lived experiences. These systems can't feel empathy or make moral judgments based on love and personal connection. They are tools that should make us smarter, not control our actions or future potential.

Systematic Nature of AI

Understanding how AI actually works should help all of us appreciate both its capabilities and its limitations better. AI's pattern-matching is based on statistics, not true understanding. If your child asks a generative AI tool to create a poem about friendship, the system is combining patterns it has learned about poetry structure, friendship vocabulary, and emotional expression. It's not actually "understanding" what friendship means to your child or drawing from any real experience of connection and loyalty. It's simply mimicking how humans have expressed the "concept" of friendship in text and examples over the years.

This is why AI responses can feel both impressively sophisticated and oddly generic simultaneously. These systems excel at identifying what's typical or common, but then can miss the wonderful, messy, unpredictable differences that make human expression special. When your teenager writes about their best friend, they're drawing from actual memories, inside jokes, and genuine emotion that no AI system could ever replicate.

The systematic approach of delivering responses to us is both AI's strength and the nature of its limitations. AI can process vast amounts of information quickly, but it can't bring the type of personal experience, creativity, and authentic understanding that we as humans naturally possess.

Importance of Celebrating Human Intelligence

Humans, of course, combine logic with creativity, emotion, and lived experience when we process information. The context we bring is also highly individual. We contribute understanding, read between the lines, and make intuitive leaps that go far beyond AI's pattern recognition capabilities. Even as we process information in similar ways to one another, we sprinkle in our own personal perspective based on individual experiences, values, beliefs, and interests. It's what makes us unique from one another, as well as wonderfully complex and unpredictable.

Consider how a human child actually learns about friendship compared to how AI "learns" about relationships. Your child experiences the joy of making a new friend, as well as the hurt of being excluded. They experience conflict between friends and then the relief of forgiveness. Our "human" intelligence comes from combining the good, the bad, the painful, and what's joyful and beautiful—and often all at once.

We need to teach our children (and remind ourselves) that empathy, moral reasoning, creative problem-solving, and understanding nuance are uniquely human qualities we should always seek to better cultivate. Humans can sense when someone needs comfort, even if they haven't explicitly said so, and navigate the unspoken rules of social situations. We can make ethical decisions that are not just logical but also reflect our values, relationships, and beliefs—and deal with any judgment or consequences that may come from making those decisions.

Power of Human Unpredictability

Humans are unpredictable in ways that make us unique and difficult to fully replicate. When children ask unexpected questions, challenge assumptions, or come up with creative solutions that don't follow logical patterns, they demonstrate the kind of thinking that sets us apart from AI. This unpredictability is actually a gift, not a flaw to be corrected.

Consider the mention I made in the preface about how often we use terms like "one-of-a-kind" to describe a person we consider extraordinary. Yet from a purely statistical viewpoint —if we're just data points for AI system training—this person would be an "outlier." In technical terms, they'd be eliminated to "normalize" a dataset. If that's how AI systems are trained, what happens to the possibility of being different? This is merely one of many unintended consequences of AI that we need to consider.

Researchers writing for the *Journal of the Association for Consumer Research* share this concern.[26] They note that AI algorithms risk "objectifying" us by "reducing the complexities of human beings into a set of quantifiable metrics, classifications, and risk scores." In other words, we end up stripped of our uniqueness to fit into algorithmic boxes for merely systematic output purposes.

As Professor Shannon Vallor writes in *The AI Mirror*, if AI is simply the sum of what data companies hold about us, what happens to our individual meaning?[27] What about the highly context-dependent moral distinctions we make about the human condition? These large questions aren't easily answered, which is why we all need to keep asking them. And there is no better time to start than right now.

Taking a Partnership Approach With Machines

The most practical solution is to focus on how to best "collaborate" with AI systems and combine both types of "thinking," without giving up on our uniqueness or worrying that AI will "replace" us. AI can handle routine tasks and information processing, while humans can and should provide creativity, wisdom, and ethical oversight. When your child uses AI to help brainstorm ideas for a school project, they're getting computational support. But when they decide which ideas align with their values, add personal perspective, and come up with creative connections that AI missed, they're exercising irreplaceable human judgment.

The goal isn't to compete with AI or avoid it entirely, but to understand how humans and AI can work together most effectively. Children who learn this balance early will be well-prepared for a future where AI capabilities continue to expand, but human wisdom, creativity, and ethical reasoning continue to be irreplaceable.

Key Chapter Takeaways

We've covered a lot of ground in this chapter. We've traced AI's decades-long ascent from an academic curiosity to an everyday companion. You now hopefully better understand why AI sometimes gets things wrong and how it can introduce bias. You see that in understanding AI terminology, we create a common language that enables important conversations to occur with family, friends, coworkers, and neighbors. And that if we avoid having these essential discussions, we risk others making key decisions about AI innovation for us.

Even after spending more than two decades in the consumer technology industry myself, my family still doesn't engage in the type of robust conversations about technological innovation that we should. We're all learning together and need to be persistent in weighing in on the issues that will shape our future and particularly that of our children.

Putting AI Innovation into Perspective

Understanding AI's long march to this moment helps us understand where we go from here. From Alan Turing's foundational questions in the 1950s to the dramatic breakthroughs that gave us ChatGPT, we see a familiar pattern of excitement, disappointment, and the gradual progress that follows. We can also see how some of these moments feel very sudden and disruptive and how that's a result of an innovation finally catching up to our individual lives.

This historical context also helps families separate genuine innovation from industry hype (and there is a lot of that right now). When we understand that today's AI has been built on decades of research and hard work, we can approach new developments with informed curiosity rather than blind enthusiasm or paralyzing fear.

Learning from AI's Limitations

Throughout this chapter, we've also explored how AI systems can produce inaccurate information through various technical shortcomings, from insufficient training data that creates cultural blind spots to the pattern-matching that can result in fabricated content. We reviewed how AI's lack of cultural context can lead to suggestions that don't align with family values and how bias operates subtly by amplifying existing societal inequalities.

Understanding these limitations helps us all approach AI systems with healthy skepticism without overlooking AI's potential. When we know that AI systems can "hallucinate" by filling knowledge gaps with fabricated information, we're better prepared to verify important claims and teach our children to do the same. When we recognize that bias arrives in the same authoritative tone as accurate information, we develop the critical thinking skills needed to question what we're hearing.

Holding on to What Makes Us Human

Perhaps most importantly, we've established that AI systems are technologies designed to enhance what we can achieve, not replace what makes us uniquely human. Our capacity for empathy, moral reasoning, creative problem-solving, and beautiful unpredictability is who we are and what makes humans special.

When your child shows genuine concern for a friend who's struggling, finds a creative solution that no adult has thought of, or asks an unexpected question that changes how the whole family thinks about an issue—these moments are what define human intelligence. As Professor Vallor notes in *The AI Mirror*, there is no algorithm that can reliably capture what's "good," "evil," or "virtuous." And we should assume there never will be. Teaching children to value and cultivate our distinctly human capabilities while leveraging AI's computational strengths creates the foundation for productive collaboration and a better future.

The Path Forward to the Future

Teaching our children to be thoughtful and confident in the face of this new technology should be our top priority. We all deserve to have a say in where we go from here and to remain active, rather than passive, participants in an AI-integrated world.

The months and years ahead will bring constant change, making it challenging to stay focused, avoid hype, and make well-informed decisions. But the principles we've established here—critical thinking, informed curiosity, and maintaining human agency—are a good first step in guiding us through whatever comes next.

The conversation we've started in this chapter will continue throughout this book as we explore AI's specific impacts on education, privacy, and family life. But first we need to explore the ways in which AI already exists in our lives and has been here for quite some time already.

Chapter 2
Where AI Exists in Your Family's Life Today

Recently, my friend Tom recounted the awe he felt when he started noticing how AI had permeated every corner of his family's life all at once. As Tom planned a day trip to visit his sister, he realized that within minutes his family had unconsciously engaged with AI at least a dozen times.

Tom's partner asked Alexa about the weather forecast while commanding their "smart" coffee maker to start brewing. His teenager used Google Maps to check traffic patterns and asked Siri to recommend a vegetarian lunch spot near her aunt's house. Tom used Alexa to quickly add items to the family's shared Amazon grocery list while his youngest played an educational, adaptive game on her iPad.

None of them had ever stopped to think that they were "using AI," he said. "Until you suggested that AI was already part of most of our lives, I thought it was simply something we had time to consider and choose to use, like ChatGPT. But it's now clear that AI has been here for some time."

The seamless integration of AI into our daily routines reveals both the technology's remarkable advancement as well as our ease in adopting new apps, tools, and platforms. We've become so accustomed to AI-powered features that they've silently faded into the background of our lives, much like how we rarely think about the electricity that powers our homes or the broadband that enables our digital activities. But as the landscape continues to shift, it's in all of our best interests to pay a bit closer attention.

Coming to Terms With Our New Reality

This invisible presence of AI creates a unique challenge as we engage in conversations at work, school, and in our communities about what role new technology should play. But the discussion shouldn't just be about generative AI but also the AI that is now integrated into all the familiar platforms we've used for years.

As families, this also means coming to terms with how our digitally native children are already far more fluent in AI than we think. For instance, consider that the Xbox 360 Kinect launched in 2010 and tracked body movements to make games interactive using early AI methods. Or that Siri is built on foundational AI technology and has been part of our lives since its 2011 launch. Understanding how these early advancements in products and services were created is essential to helping us make informed decisions about privacy, safety, and appropriate use today.

The fact is, we've developed a deep comfort with foundational AI technologies for at least a decade. It's been quietly shaping everything from the news we see in our social media feeds to the routes we drive and the movies we watch. By taking inventory of AI's current presence in our homes, schools, and communities, we can be far more active drivers of our family's AI journey. This empowered, active awareness will also be the basis from which everything that follows flows.

What You Will Discover in This Chapter

In this chapter, we'll map out the AI landscape that already exists around us. I'll help you identify how AI technology has been quietly working behind the scenes and why it's now become a part of our routine experiences.

Building on the foundational vocabulary we established in Chapter One, we'll explore where AI currently operates in your home, from smart devices to entertainment systems. We'll look at how to consume news and information about AI while putting the "hype" into perspective. I'll also explain the current regulatory environment, what AI protections are currently available to families, and how AI is impacting careers.

As most families know, AI is also already in schools. But because that is such an important topic, I dedicate Chapter Three to considering its current impact in the classroom.

Understanding AI's presence right now requires looking beyond the obvious chatbots and voice assistants. We need to recognize the sophisticated systems already powering everything, from how we commute and work to the way we interact with friends and choose our entertainment. In this chapter, I'll help you distinguish between new capabilities like generative AI and the AI-enhanced versions of familiar technologies such as smart devices, voice-activated services, and online shopping recommendation engines.

By the end of this chapter, you'll have a clearer picture of AI's current role in your world and a framework for evaluating new developments as they emerge. I hope this foundation will enable you to move from reactive responses to generative AI platforms to more proactive awareness of all the AI technologies shaping our families' lives today. So let's get started by looking at all the surprising ways AI is all around us right now.

Familiar AI Touchpoints

As you now understand, the invisible infrastructure supporting most of our digital experiences is vast and increasingly powered by AI. For instance, shopping search sites employ AI to understand product browsing. Cloud storage services use AI to organize your family photos and help you find that picture from last summer's vacation. Even email platforms rely on these systems to protect against cyber threats.

These different "foundational" AI technologies, as we covered in Chapter One definitions, touch nearly every digital interaction we have, and often without our conscious awareness. Understanding this larger context can help families move from unconscious consumption of technology to intentional engagement with the digital world. So let's review the platforms, tools, and infrastructure we may not have previously understood to be under the umbrella of AI.

The Home's Hidden AI Network

Voice assistants represent the most obvious AI presence in many homes today, but even these familiar devices do far more than we realize. When we ask Alexa to play music or control smart lights, we're engaging with natural language processing that interprets speech, identifies intent, and coordinates responses across multiple systems. It's like having a translator who understands what you're saying and figures out what you actually want and then manages to make it happen across your entire home.

These systems use machine learning to adapt to individual

speech patterns and preferences over time, which explains why Alexa might better understand your teenager's requests than your voice, or vice versa. Your family is essentially training the AI to work better for each of you through daily interactions.

Smart home devices extend this AI presence throughout our living spaces in ways we rarely notice. AI-enabled thermostats learn your family's schedule to optimize energy usage automatically. Think of it like having an invisible household manager who pays attention to when everyone's home and adjusts accordingly.

Modern security cameras use computer vision to distinguish between family members, pets, and strangers, reducing those 3 a.m. false alarms when the cat walks by the door. Smart doorbells employ facial recognition technology to identify frequent visitors and can announce specific people's arrivals (making it handy to know if it's an Amazon package delivery or your neighbor).

These features bring remarkable convenience and can provide life-changing autonomy for family members who previously needed assistance with daily tasks as well. A grandparent living independently can control their entire home environment through voice commands, while parents can monitor kids and adjust home systems remotely during the workday.

But here's what we can't overlook: smart home networks also bring trade-offs to our privacy and may introduce additional risk. Because, of course, every AI-powered device collects data about your family's routines, preferences, and behaviors. This information often travels to company servers for processing and improvement, creating potential privacy and security vulnerabilities that affect everyone in the household. The goal isn't to avoid the technology, as this data is the basis for how these systems operate, but we need to understand the trade-offs. It also shows how important it is to use a company you trust and that you are clear about its terms and conditions and privacy policy too.

AI-Powered Entertainment

Home automation represents just one layer of AI's invisible presence. Perhaps even more pervasive is how AI shapes what we watch, listen to, and discover during our downtime. We frequently don't realize how extensively our entertainment choices have been influenced by AI recommendation systems for years.

That said, today's streaming-based data analysis is exponentially more sophisticated. Everything from pause patterns and browsing behavior to time spent reading descriptions feeds into systems that predict what your family wants to watch. It's like having someone who's been secretly watching your viewing habits for years suddenly become your personal entertainment curator.

Every major platform employs similar data analysis tactics, often intersecting with the other apps you use and data you produce daily. Music platforms like Spotify and Apple Music create personalized experiences by analyzing listening habits, while platforms like YouTube determine which videos appear in your children's feeds based on complex engagement predictions.

For families, this means that what your children watch, listen to, and discover online isn't random but carefully curated based on data about their digital behavior. Understanding this invisible influence helps families make more informed decisions about screen time, content exposure, and digital media consumption at home.

These recommendation systems also shape far more than entertainment choices. They influence our family's exposure to information, ideas, and cultural content while informing what type of new content gets produced—including the increasingly AI-generated entertainment we might see in the future.

The AI Behind Everyday Tasks

Beyond entertainment choices, AI has also quietly taken over many of the routine digital tasks we perform throughout the day, often in ways that save us time without us even realizing it. Email systems use AI extensively for spam filtering, priority organization, and those suggested responses that save us time throughout the day. It's like having an administrative assistant who knows your communication style and handles the routine stuff so you can focus on what matters.

Smartphones employ predictive text algorithms that catch typos, learn your writing style, and make phrase recommendations (which is why your phone might suggest different words than your teenager's device). Translation apps use neural networks that process information similarly to how the human brain works, enabling real-time communication across language barriers—basically turning your phone into a universal translator.

GPS navigation represents one of AI's most practical daily applications for families. Maps don't just calculate the shortest route but use machine learning to predict traffic patterns, construction delays, and parking availability. This means the route suggestion for your morning school drop-off considers real-time conditions that a simple distance calculation would miss. For families with newer vehicles, AI systems monitor driving behavior, predict maintenance needs, and assist with parking, creating a more integrated travel experience.

The Invisible Influence on Our Wallets

This invisible AI assistance extends beyond navigation into areas that directly affect families' financial decisions and spending patterns. Online shopping platforms use machine learning to personalize product recommendations, optimize search results, and detect fraudulent transactions. When you shop online, AI foundational technology analyzes your browsing history, purchase patterns, and even how long you spend looking at products. It's like having a salesperson who's been studying your preferences for months, except this salesperson can now influence thousands of customers simultaneously.

You might also be surprised at how the AI technologies can introduce dynamic pricing too. Algorithms often adjust costs in real-time based on demand, your location, and even your device type, which means the price you see might differ from what your neighbor sees for the same item for everything from homewares to flight tickets. Similar to the "personalization bias" I described in Chapter One, you might think you're actively choosing products, but more frequently just being influenced by algorithms designed to maximize purchases.

AI also shapes how families pay and manage finances. Credit card and banking AI systems handle fraud detection, spending analysis, and financial planning suggestions. Banks use machine learning to identify unusual spending patterns, offer new cards during key shopping periods, and suggest loan rates when families tend to carry more debt.

Understanding these AI influences helps explain why certain products appear repeatedly in your feeds, why prices seem to fluctuate, and how financial institutions time their offers. This awareness enables more intentional decision-making about both spending and the personal data you share through these platforms.

Thinking Through Our Most Sensitive Data

AI's data collection becomes even more personal when it comes to a family's health and wellness information. Fitness trackers and health apps use AI to analyze movement patterns, sleep quality, and heart rate data to provide personalized wellness recommendations. Devices learn individual patterns over time, which is why a fitness tracker might recommend different activity goals for each family member or alert one person about irregular sleep patterns while remaining silent for another.

These AI health tools provide valuable insights that were previously available only through professional medical consultations. They can identify concerning health trends, encourage positive behaviors through personalized coaching, and offer mental health support to families who might not otherwise have access to these services.

But here's where it gets tricky: health information represents some of our most sensitive data. When families use AI-powered health tools, they're potentially sharing intimate details about physical conditions, mental health struggles, and behavioral patterns. This information often travels to company servers for analysis and improvement, raising important questions about who has access to your family's health data and how it might be used beyond its intended purpose.

As I'll explore in later chapters, existing privacy laws may not adequately protect families from potential bias, discrimination, or other harms from companies that profit from personalized health data. So it's something to keep on our radars.

Newer AI Tools Now in the Mix

While all these foundational AI systems have been working behind the scenes for years, the recent arrival of generative AI has brought the technology into the spotlight in entirely new ways. While generative AI has captured our recent attention and imagination, we must not lose sight of the larger picture. But the good news is that in understanding this larger land-scape, we also have more context with which to evaluate generative AI systems, such as ChatGPT.

Explosive Growth in New AI Systems

While new AI systems such as ChatGPT, Claude, and Google Gemini are the most talked about right now (mostly for their singular chatbot interface), generative AI functionality is being integrated into many of the existing platforms we use each day. And while the biggest generative AI systems are often in the news for their missteps, some newer integrations have problems that shouldn't slip from our radar either.

For instance, in January 2025, the Federal Trade Commission (FTC) announced that it would refer a complaint to the Department of Justice (DOJ) alleging that Snap's "My AI" chatbot posed a risk to kids.[1] While this statement was shared in the interest of "keeping the public informed," it was without any detail (or followup at the time this book was published). But a lack of detail shouldn't mean the news goes unnoticed. Considering Snapchat's popularity with kids, it's an unsettling development.

The point is that we need to keep our eye on all updates and news related to AI tools and the implications for our kids.

And that starts with understanding the newest systems and those familiar ones that are getting generative AI updates on a rolling basis.

AI Integrations into Familiar Platforms

Many familiar educational platforms are developing AI-specific systems or partnering with large AI companies to integrate technology into existing platforms. Khan Academy's Khanmigo, for instance, provides personalized tutoring integrated into their established educational products. Microsoft's Copilot has been introduced into many school systems through existing educational contracts, while Google Gemini works within Google Workspace for Education environments that many schools already use.

These educational integrations often include enhanced privacy protections, activity monitoring for teachers and families, and content specifically designed to support rather than replace traditional learning approaches. But it can be difficult to identify which features use AI or to understand the contracts and data-sharing agreements that schools have made with these vendors.

Your child might be using AI-powered tools at school without you realizing it as well. And the data from those interactions may be used to improve the AI systems or inform educational decisions in ways you haven't explicitly consented to.

This doesn't mean educational AI is inherently problematic, but it does mean families need to ask more questions about what AI tools their schools are using and how student data is being handled. Starting conversations with teachers and administrators about AI policies can help ensure you're informed about your child's educational AI experiences (something we'll cover in depth in later chapters).

Social Media AI Chatbots

Every major social media platform uses sophisticated AI algorithms to decide what content appears in your child's feed, and it's been happening for years. What's particularly notable now is how quickly these platforms have integrated AI chatbots, with AI chat features becoming part of everything from X (Grok) to Snapchat (as mentioned) and the Meta family of apps.

And while these chatbots might seem similar to ChatGPT, social media AI systems have access to much more personal data. When your teenager chats with Instagram's AI, the system knows their posting history, direct messages, search behavior, and social connections, creating incredibly personalized but potentially privacy-concerning responses.

While this creates potentially more engaging conversations, it also means these systems can build detailed profiles of your children's thoughts, concerns, and social interactions. They might know when your teenager is feeling anxious about an upcoming test, excited about a crush, or struggling with friend drama. We can forget how deeply personal and private this information is, yet how it's being shared with social media platforms regularly.

It's important to remember that social media AI chatbots are designed primarily to increase engagement and time spent on platforms, not to provide educational value or emotional support. That's not necessarily an indictment of these platforms, but it's an essential distinction when considering their appropriateness for our families. We also shouldn't give a platform a pass just because we use it already. As in the case of the

FTC news about Snapchat, we need to evaluate everything on an ongoing basis.

The Wave of New Specialized Offerings

Beyond the mainstream and educational platforms, a rapidly expanding ecosystem of specialized AI tools is emerging. These include AI companions designed for emotional support, AI tutors focused on specific subjects, creative AI platforms for art and music, and even AI systems designed to simulate relationships or provide therapy-like conversations.

Some of these specialized platforms offer genuine value; for instance, an AI language learning tutor that adapts to your child's pace or a creative writing AI that helps spark imagination. But still others exist primarily to collect data, build engagement, or profit from subscription fees without providing a clear benefit to users.

The challenge for families is that some specialized platforms have less oversight, fewer safety measures, and unclear privacy policies compared to mainstream alternatives. They might not have the resources or incentives to implement the same safety features that companies like OpenAI, Anthropic, Microsoft, or Google have developed.

When evaluating specialized AI platforms, families should first seek to understand the company behind the technology, what data the company collects, and how they generate revenue. It's also important to understand any age restrictions or information shared with third parties. If your questions aren't met with clear, satisfactory answers, then it's usually a red flag and a platform to avoid.

Why AI Platform Distinction Matters

Recognizing different types of AI systems helps families make more informed decisions about technology use. When your teenager wants to try a new AI platform, you'll be better able to determine whether it's a mainstream generative AI platform with educational potential or one designed primarily for entertainment that maximizes engagement.

These distinctions help families set appropriate expectations and boundaries. Conversations about using ChatGPT for homework help should differ from discussions about social media algorithms or AI companion platforms. Understanding these categories provides a framework for evaluating any new AI tool your family encounters.

The key is developing a family philosophy about AI use that considers both the benefits and risks of different platform types while maintaining open communication channels to discuss any family concerns. We're all learning to navigate this landscape together, and there's no perfect roadmap. But by understanding the different types of AI systems and their various motivations, we can make more intentional choices about the technology that shapes our family's daily experience.

The goal isn't to avoid all AI innovation but to approach it with the same thoughtful consideration we'd give to any other significant influence in our children's lives. Understanding the facts, industry landscape, and specific family rules and values when it comes to AI also makes evaluating any news around the technology easier to sort through and consider. That is to say, separating the truth from "hype."

How "Hype" Affects Our Understanding of AI

While you may have missed the simple but important Snapchat news, you've probably obsessed over one of the more dramatic headlines circulating about—and that's a problem. Because we get so caught up in our emotional reactions to "clickbait" that we can overlook the dry but critical information that we actually need.

Depending on your preferred news outlet, AI is either a revolutionary transformation or a step toward complete societal collapse. This type of coverage leaves families feeling either terrified or unrealistically excited, but rarely just better informed. Compounding the issue is the way in which AI systems are both the subject of news coverage as well as the actual technology "powering" coverage that is algorithmically tailored. And worse, tailored in a way to elicit the most emotional response possible so we click, read, and then media generate revenue.

But when we start to understand why extreme AI coverage dominates the news and work to evaluate content with a cooler head, we start to make better decisions about the future. We can look around, see where AI exists in our lives, and then consider whether something represents genuine change or is just meaningless hype.

Economics of Igniting Emotional Responses

News about AI follows the same playbook as every other piece of digital content fighting for your attention. Dramatic headlines generate clicks, clicks generate revenue, and revenue drives even more dramatic headlines. When a news outlet publishes a story that elicits an emotional response in us, that story performs far better than a measured analysis like, "New

AI Tool Shows Promise but Has Notable Limitations." The reasonable headline doesn't trigger the emotional response that drives people to share articles with their worried friends or excited colleagues.

Unfortunately, both wild optimism and dire pessimism often serve the same financial interests. Venture capital firms have invested billions in AI startups, creating enormous pressure to demonstrate the technology's "revolutionary" capabilities. At the same time, authors and consultants build entire careers around AI warnings, profiting from the very anxiety their predictions create.

Social media algorithms worsen this by promoting content that generates the strongest reactions. A balanced post about AI's educational benefits and limitations might receive modest engagement, while a post claiming "AI Is Destroying Our Children's Minds" spreads rapidly, reaching thousands of concerned families within hours.

To top it off, even the most trusted news sources tend to cover AI from either an overly optimistic or disproportionately pessimistic standpoint. Researchers analyzing news coverage of AI in twelve countries even found a consistent lean to the coverage depending on region.[2] Newspapers such as the *The New York Times* and *The Guardian* (UK) tended to frame AI in negative terms, while articles in China Daily and Bangkok Post generally covered AI in a more positive light.

What can be difficult for families is that we are in a place right now where we are both learning how to use AI technology and also trying to figure out what we want to think about it all. Too many negative stories about AI may affect our willingness to experiment with these tools, while overly glowing coverage may cause us to reject AI as a disappointment when it doesn't deliver. We need balance, and it's on us to seek it out.

How to Spot Unreliable AI Information

Learning to distinguish between dramatic content designed to trigger an emotional response and more balanced news is becoming an increasingly important media literacy skill for families. But there are some easy things to look for when evaluating a story, and it can start to become automatic to do.

RED FLAG HEADLINES

You can often spot questionable coverage by its headline. With words like "shocking," "revolutionary," "destroys," or "changes everything," you'll know immediately that you should consider the detail carefully. Quality AI coverage (and arguably any coverage) uses more measured language and acknowledges complexity rather than promising simple solutions to complicated issues.

UNREALISTIC TIMEFRAMES

Stressfully short and imminent timelines are also typically a giveaway. Stories that promise dramatic changes within months or make broad predictions about what will happen "by 2028" typically reflect speculation rather than informed analysis. Real technological change usually happens more gradually and unevenly than timeframes suggest.

SINGLE-SOURCE STORIES

Coverage that only quotes a single source, and especially industry insiders or researchers with clear financial stakes in AI development, typically lacks the balanced perspective families need. Quality coverage includes multiple expert voices, acknowledges uncertainty, and presents different viewpoints on complex issues.

MISSING CONTEXT IN THE DETAIL

Often the most misleading articles are those that share research or news out of context. For instance, stories that present AI developments without explaining current limitations or hammer implementation challenges without mentioning how they will improve with time. These stories often omit the practical information that families might find useful in trying to put the pieces together.

Also look for stories with specific examples rather than vague generalities. Good content provides practical guidance around current decisions rather than speculative future considerations. Look for quotes from a diverse array of experts, including educators, ethicists, and independent researchers.

Importance of Tuning Out Information Overload

The AI hype cycle provides an excellent opportunity to talk about media coverage with our kids. When families encounter AI news together, it can help to ask questions like, "Who benefits if we believe this story?" "What evidence supports these claims?" and "What perspectives might be missing from this coverage?"

Children can learn to distinguish between news reporting and opinion content, identify potential conflicts of interest, and seek multiple sources before forming conclusions about AI developments. These skills are valuable regardless of the topic.

Most importantly, remember that we are modeling information literacy for our children every day. When kids see us approaching AI news with curiosity rather than panic or blind enthusiasm, they develop confidence in their ability to evaluate complex information and make thoughtful decisions.

Think of it like parenting itself. You don't need to solve every possible future problem your children might face, but you can help them develop the skills and judgment to handle challenges as they arise. The same approach works for navigating AI news—focus on building understanding and critical thinking skills rather than trying to predict or control an unknowable future.

A measured approach to consuming news becomes even more effective when we understand what AI policies and protections are actually in place. Having that context is just one more piece of the puzzle that helps us feel confident about what's really happening versus what the headlines want us to believe.

AI-Specific Regulations and Protections

If you've been wondering who is responsible for protecting our families from any potential AI harms, it's a good question. Listening to the AI-related legislative discourse around the country feels like trying to follow traffic rules in a city where half the street signs are missing, the other half contradict each other, and new roads keep appearing overnight.

Governments worldwide are working to balance innovation with consumer protections, especially for children and families. But governing AI innovation comes with a double challenge. First, AI isn't one thing but a suite of technologies, as we discussed. Second, while we have existing laws that touch on important aspects of AI use—like data privacy laws or those protecting kids' data in schools—these weren't designed with AI in mind. Even our strongest protections are often unable to address how AI systems are actually built and operated.

On top of it all, the intersection of federal, state, local, and district policies creates a complex web of overlapping rules and gaps in protections, making an already difficult situation even more challenging. But understanding this regulatory maze can help families navigate AI innovation more safely while also providing opportunities to get involved in shaping its future use. We'll look first at some of the protections that are as close to "AI-specific protections" as we can get. Then in future chapters we'll dive more into the topic of data privacy at home and in schools.

Fuzzy Detail at the Federal Level

President Trump's April 2025 Executive Order, Advancing Artificial Intelligence Education for American Youth, represents the most significant federal AI educational mandate to date.[3] It establishes a White House Task Force on AI Education and directs federal agencies to prioritize AI integration in K–12 schools.

It sounds impressive, but with a few important caveats. While the order sets ambitious goals for AI education, it doesn't specify concrete timelines for implementation and leaves much of the detail to federal agencies and local entities. Executive orders set the direction and priorities but generally don't include the tactical details that, in this case, translate directly into classrooms. This creates a real challenge given that many educators are still catching up on AI themselves.

Related to schools and teacher preparation, the Department of Education has been directed to prioritize AI in teacher training grants and provide guidance emphasizing transparency, equity, and student privacy in school AI use.[4] And in terms of protections, the FTC has increased enforcement of existing consumer protection laws against AI companies, particularly targeting deceptive practices and algorithmic bias.[5] But despite these efforts, much of the onus for specifically protecting kids when it comes to AI still rests with state and local governments.

Disconnected Puzzle of State and Local Policy

Without federal clarity and consistency, states, cities, and school districts have created their own AI approaches with wildly different outcomes. Some states are proposing restrictions on how AI companies collect children's educational data; others are implementing AI auditing requirements for hiring; and many are introducing bills requiring AI literacy in computer science curricula. In our boundaryless digital world, enforcing these varied laws and protections can be nearly impossible, leaving families frustrated and confused.

School districts nationwide are also developing their own AI policies with dramatically different approaches. Some ban AI systems entirely; others have embraced AI innovation with detailed guidelines, while many remain uncertain about where to begin. This inconsistency means families in neighboring districts may face entirely different AI educational experiences and protections.

As we'll cover in more detail later, when districts also "ban AI," they're typically targeting only generative AI like ChatGPT while overlooking the far greater landscape of AI functionality already well integrated into their systems. The Digital Education Council, a consortium of more than 100 institutions, shares this sentiment. In a recent report, they noted that by focusing only on generative AI, schools are taking too narrow and reactive an approach.[6] The risk is that schools not only miss the larger AI-driven picture, but they also miss any innovation they find problematic that sneaks up right behind generative AI.

International Policy Context to Consider

The EU's AI Act, enacted in 2024, represents the world's most comprehensive attempt at AI regulation.[7] While US families aren't directly subject to EU law, many AI companies may adopt EU standards globally (for ease of compliance), potentially extending some protections to American families. However, it's too early to assess whether these regulations effectively protect users or whether they might hinder beneficial AI innovation.

Countries like Singapore and Canada have emphasized transparency and accountability in their emerging AI regulations, while China has implemented restrictions on AI algorithms affecting children on social media platforms. Yet these approaches create their own complexities, such as the contradiction between China's domestic AI restrictions and how Chinese companies like TikTok handle foreign children's data.[8]

These international efforts illustrate different regulatory philosophies that American policymakers might consider, but families should understand that we don't yet know which approaches will prove most effective. The challenge for any regulatory framework is balancing consumer protection with the innovation that drives technological advancement and economic competitiveness.

Children-Specific Regulations

Existing children's privacy laws like the Children's Online Privacy Protection Act (COPPA) technically cover AI, but enforcement remains challenging.[9] While COPPA requires parental consent before collecting data from children under 13, determining what constitutes "data collection" by AI systems is nearly impossible to define and control in practice.

Proposed federal legislation such as the Kids Online Safety Act (KOSA) would provide additional protections for minors regarding AI-powered social media and educational tools.[10] However, as of this writing, the legislation has stalled due to concerns from a wide range of parties. Specifically, technology companies have raised issues regarding the bill's overly punitive provisions. And equally, advocates, including the American Civil Liberties Union (ACLU), have objected to the legislation because it could restrict free speech, particularly for marginalized communities of kids that rely on online platforms for support and information.[11]

While many of these regulatory proposals address important issues like algorithmic manipulation and age-appropriate content filtering, even when passed, such regulations would likely be difficult to enforce and may not keep pace with the rapid speed of AI innovation. As we'll cover later in the book, data privacy laws, particularly those that relate to schools (and including COPPA), should be able to address, in theory, some of the more urgent needs related to AI's potential data privacy violations. But again, these laws struggle to match the reality of how AI technologies actually work.

Industry Attempts at Self-Regulation

Many AI companies have implemented privacy policies that go beyond current legal requirements, but these voluntary measures often miss fundamental connections between data collection and AI training. Privacy policies might focus on transparency about data collection while using that same data extensively for their AI development, all without independent oversight.

Most large companies, organizations, universities, and nonprofits recognize that AI cannot grow and thrive without ethics at the core of AI development. Not only will consumers increasingly reject industry advancements that feel unsafe and predatory, but governments will intervene with regulations that may make innovation more difficult.

Industry-backed organizations such as Partnership on AI are attracting diverse voices to help guide this effort.[12] And while it's a win-win approach for everyone to establish data privacy and ethical frameworks voluntarily at the industry level, families shouldn't rely solely on company good intentions to protect their children's best interests.

Why Our Voices Matter to Policy Discussions

A fragmented regulatory landscape puts more responsibility on families to protect themselves, but it also means families' voices carry more weight in important discussions. When federal guidance is vague and state policies are inconsistent, local school boards and district administrators often look to families to help develop solutions.

Engaging with AI regulation doesn't require technical expertise, just family-specific perspectives on privacy, safety, and educational value. When families share real experiences about AI's impact on homework routines, learning outcomes, and digital safety, they provide essential context that policymakers need when developing regulations. This practical feedback strengthens any policy or rules that emerge.

Your involvement can take many forms, from attending school board meetings to commenting on proposed district AI policies to contacting representatives about federal legislation. The key is recognizing that this regulatory landscape is still being written, and families have more influence over the outcome than they might realize.

How AI is Disrupting Work Today

While we're trying to figure out the regulatory landscape together, there's another domain where AI's impact is already creating waves: work. What's notable about AI's impact on jobs right now is how older adults in a given household are often experiencing the same career uncertainty and stress as their college-age kids.

Compounding this is the fact that much of the disruption in the workplace is not just due to AI but also the "idea" of AI right now. Sometimes employers have even been too quick to embrace AI "efficiency," laying people off only to hire them back again.[13] While in other cases employers have been slow to hire until they know the part AI will play or have limited the hiring of entry-level roles.

On the positive side, there are, of course, many examples of AI already enhancing careers. My friend Priya, a radiologist, initially feared that AI image analysis might eliminate her profession. And while she remains unsure about the long-term impact of AI on her work, the early signs have been surprisingly positive.

The AI systems available so far have allowed her more time to focus on patient consultation and diagnostic decisions. "I think my fear of AI replacing my work has not been fully rational," Priya explained. "I actually now see the potential it has for handling routine analysis so I can spend more time on the work that requires human judgment, empathy, and problem-solving."

Priya's experience also aligns with recently published research about AI's impact specifically on radiology. The authors found that AI innovation has been a positive advancement for the field but also very user-dependent.[14] That is to say, it is a beneficial tool that still requires human expertise to produce the best results.

But overall, by understanding AI's specific capabilities and limitations, we can help our kids sort through the uncertainty of future work while also finding some measure of sanity in looking at our own job prospects at the same time.

The Rocky Job Outlook Ahead

Just because AI may ultimately have a positive impact on careers doesn't mean that we aren't in for some instability while companies figure it all out. And for recent college graduates, a troubling picture has already emerged—particularly concerning since we need this generation to lead us into tomorrow while we're simultaneously worried about our own job security. According to the Federal Reserve Bank of New York, the unemployment rate for recent college graduates in June 2025 was 4.8%, compared to 2.7% for all workers with a college degree.[15] If you have a young adult at home, you are likely well aware of this unfortunate picture. It's a worrying trend that analysts and media are indeed attributing to, in part, AI displacement in the workplace.[16]

Generation X parents, particularly those in creative fields such as film, journalism, or advertising, have also seen their areas of expertise shrink already.[17] But for these parents, the ultimate fear is that they may never work again. It's something my good friend, Executive and Certified Mentor Coach Tracy Irvine, has seen recently with her clients.[18] And the best way to most productively address the situation, Tracy says, is to become as knowledgeable as possible about AI, while at the same time taking a good look at your field and deciding if you should pivot careers. "It can be a fantastic opportunity to take a fresh look at one's career, pursue any new certifications, or even consider training for a new career," she said. "It's never too late to make the switch, and AI might just provide the necessary incentive to do this."

None of this means that the future of work is fully bleak. Every indication points to AI ultimately introducing opportunity, but for many sectors it's been a tumultuous start. Until it all shakes out, there will be career sectors and certain demographics that find the transition particularly difficult—and families will need to navigate this transition with patience and adaptability.

Think Job Transformation, Not Elimination

History does show us that technological advancements are more apt to change professions than eliminate them. It's like when calculators became common in the 1970s. They didn't eliminate the need for mathematicians but instead changed what the work looked like and freed up professionals to focus on more complex problem-solving. Or how word processors didn't eliminate writers but changed how writing gets done, making it far more efficient.

There will be jobs that disappear because of the efficiencies that AI will bring, but there will be others that sprout up in their place. And to secure those opportunities will require a focus on acquiring new skills. According to the World Economic Forum's *The Future of Jobs Report 2025*, AI will create demand for roles that require human oversight, creativity, and ethical judgment in working with AI technology.[19] The study also noted that the skills needed to transition to future industries include resilience, flexibility, agility, and technological literacy. Interestingly, these are all highly "emotional" skills that many of us already possess and just need to better highlight to employers, which I'll cover in more detail in later chapters.

But overall, our children will be well-positioned to enter the job market as it continues to evolve if they stay flexible and broaden their skillsets and knowledge base. And by the same logic, adults should also benefit from rethinking what they can offer and embracing the same flexibility mindset.

AI Can Be a Career Enhancer

There's movement and opportunity in every profession—it's just a matter of how quickly these changes will happen and how well any of us can adapt.

Healthcare professionals are already using AI for diagnostic assistance, treatment planning, and patient monitoring, which frees them up to spend more meaningful time with patients. Teachers and educators are integrating AI for personalized instruction, assessment, and administrative tasks so they can focus their energy on what drew them to teaching in the first place: sparking curiosity and connecting with students.

Even skilled tradespeople are benefiting from AI-powered tools for design, diagnostics, and efficiency optimization. But they're still the ones with the irreplaceable hands-on skills and customer relationships that make their work valuable.

The pattern here is clear: AI handles routine tasks while humans focus on what we do best—creativity, empathy, problem-solving, and building relationships. Understanding how jobs will evolve should help families prepare kids for these shifts and the types of jobs they might want to pursue someday.

New Job Categories Emerging Daily

There are also entirely new job categories emerging that didn't exist just a few years ago. These roles require a mix of traditional skills with AI-specific expertise, making them perfect for recent graduates or older adults looking to change careers.

AI ethics and governance specialists will help organizations use AI responsibly, ensuring that AI systems are fair, transparent, and aligned with human values. Human-AI collaboration designers will focus on optimizing how humans and AI systems work together most effectively.

Since data is so core to AI creation and growth, we'll need AI training and data specialists who ensure AI systems learn from high-quality, representative data while protecting consumer privacy and preventing bias. Think of them as librarians for the digital age, making sure AI systems learn from the right sources. We'll also need safety and security professionals to protect AI systems from misuse while ensuring they operate safely and reliably.

Finally, AI-enhanced service professionals will work in fields like mental health, education, and customer service, where AI systems enhance human capabilities but cannot replace the essential human element of care and understanding. These are only a few examples, but the point is that there will be so many new opportunities—we just need to be flexible and ready to grab them.

Preparing for Work Uncertainty Together

Perhaps the most important lesson for families is that we cannot predict exactly how AI will develop or which specific careers will emerge. The intersection of all of this uncertainty happening all at once with multiple family members can also be emotionally challenging. It's like everyone in the house is learning to drive at the same time, but the roads keep changing and no one has a reliable map.

But kids who can think critically about technology's role in society and who maintain curiosity and adaptability will be more than prepared for whatever comes next. And as we navigate this new world together, we all benefit from diversifying our skills and building a resilient mindset that will enable success regardless of how AI technology continues to evolve.

Key Chapter Takeaways

By now, your family should understand that AI isn't some futuristic technology slow to arrive but instead something already deeply embedded in our daily lives and routines. From the moment we check our smartphones in the morning to turning on a streaming service at night, AI systems are working behind the scenes to personalize, predict, and optimize our digital experiences.

As my friend Tom realized that morning when he started to pay attention to his family's digital activities—recognizing the existing AI touchpoints all around us will give us the confidence and agency to tackle what's next.

AI's Hidden Presence Around Us

I hope you also now see how extensively AI is already woven into the fabric of our lives. In our homes, voice assistants coordinate smart devices while recommendation algorithms curate our entertainment choices. Our cars navigate using AI-powered GPS systems that predict traffic patterns, while our children's educational apps adapt to their individual learning styles through machine learning.

Even routine tasks like email organization, online shopping, and financial management rely heavily on AI systems working invisibly in the background. This pervasive presence means that the question for families isn't whether to engage with AI, but how to engage thoughtfully and intentionally.

The smart home devices, streaming platforms, and educational technologies your family already uses have established a foundation of AI interaction that we can build upon as more sophisticated systems become available. Understanding these foundations also helps us feel more confident about navigating the newer AI tools that are emerging regularly and now dominating the conversations about AI.

Media Hype vs. Sober Reality

Perhaps one of the most crucial points we've explored is how dramatic stories about AI often overshadow the mundane but necessary complexities of issues like the regulatory environment seeking to address these challenges. The breathless headlines about AI's dangers or promises can make it feel like we're careening toward an uncontrolled future, but the reality is far more nuanced and requires thoughtful attention.

There has never been a more crucial moment to stay focused and clear-headed about what's happening around us. The regulatory gaps we've identified around children's privacy, platform accountability, and educational AI use aren't permanent obstacles but rather areas where family voices can make a significant impact.

This awareness puts families in a position of informed influence rather than reactive confusion. You're not waiting for someone else to figure out the rules—you're understanding enough to help shape them. And simply "doomscrolling" through shocking AI headlines doesn't do much to help.

Real-Time Career Transformation

We've also examined in this chapter the current career landscape with all of its genuine uncertainty and emerging opportunities. Just as my radiologist friend discovered that AI enhanced rather than replaced her diagnostic work, most professions are being reshaped rather than eliminated. But that transformation won't occur in a straight line, and we are already seeing massive disruption to our careers.

The unemployment struggles facing both recent college graduates and their Generation X parents also remind us that this transition isn't happening in some distant future, but right here and now. Understanding this dual pressure also helps us approach career conversations with our kids as something we are experiencing together.

Forging Ahead With Realistic Expectations

What's most striking about this moment is how much knowledge and strength we already have in hand as we face a future with machines. We are longtime users and even early adopters of the technologies making up the foundations of today's AI. And all of us are capable of making more deliberate choices going forward regarding newer applications.

This isn't about having all the answers, of course, because none of us do. It's about building a knowledge base that allows our families to ask better questions, make more informed decisions, and contribute meaningfully to the conversations that will shape AI's role in society. To that end, the goal of this chapter has been to ground you in the current reality of AI so that we can tackle the more complex challenges ahead.

Now that you understand where AI already exists in your family's life and its broader implications, we'll round out Part One of the book by examining one of the most immediate areas where AI is creating both opportunity and concern: school. The classroom is quickly becoming ground zero for many of the AI discussions that matter most to families, and understanding what's happening there will help you support your children and their teachers.

Chapter 3
How AI is Already Transforming Education

Have you recently tried to help your child with their math homework, only to discover that learning fractions bears no resemblance to what Mr. Esposito taught you in the 3rd grade? Or maybe you've had to do a Google search to figure out the difference between "Singapore Math" and "Common Core." For many families, the idea that our children are now coming home and asking about yet another subject we're unfamiliar with is quite stressful.

While teaching methods have continually evolved, AI is entering the classroom at unprecedented speed, accelerating changes to education beyond anything we've seen before. Your teenager might casually mention that they're "prompting Chat-GPT" for help with an essay outline, while your middle schooler excitedly describes using AI to create a movie. Meanwhile, we're in the thick of trying to figure out ourselves how and when to use AI tools at work and home.

As my friend Damien said to me recently, "There is no way I can evaluate the use of a tool in the classroom or have an opinion about my child's work when we're talking about a technology I haven't even used yet." His frustration captures what many families are feeling right now. We're unsettled by the disconnect between AI's promise and our understanding as families of how it's being taught in schools. According to the Samsung *Solve for Tomorrow* AI survey, 88% of parents believe that knowledge of AI will be crucial for their child's future education and career.[1] But at the same time, 81% have no idea if AI is being used in schools or even doubt that it's used at all.

Remote Learning's Impact on Today's Classroom

Of course, the issue didn't begin with ChatGPT but rather, in part, when families and schools were forced into remote learning during COVID-19. According to a report by EY Parthenon, teachers use of digital instructional materials jumped from 28% before the pandemic to 52% by 2022.[2] Even though teachers are more comfortable now integrating technology into the classroom, many felt that the pandemic-era rollout was haphazard and unstructured. The sharp increase in digital learning tools during this time also means today AI can more easily show up via simple feature updates to existing platforms.

What makes this current moment even more challenging is the pressure it's heaping on not just students and families but the entire educational system. And yet, despite all of this, the opportunities for integrating AI into schools remain compelling. AI systems aren't just able to personalize learning in new ways for all students, but can better engage diverse learners, support language learning, and even provide tutoring for kids in under-resourced schools.

It's like having a teaching assistant who never gets tired, speaks every language, and can explain the same concept in dozens of different ways until it clicks for each individual student. When used by teachers who have had the time and training to consider the scope of what's possible, it's exciting. But we need to support teachers' professional development to get us there.

What You Will Discover in This Chapter

This chapter will take you on a journey through the ways in which AI is already reshaping schools and our children's educational experiences today. We'll start by looking at how AI can create truly individualized learning via adaptive "tutoring" that adjusts to a child's individual needs. From there, we'll explore how AI can open doors for kids by making complex concepts accessible through interactive simulations and personalized explanations. And we'll examine how schools are beginning to measure learning progress differently when AI is involved— moving beyond traditional grades to real-time insights about how a child actually processes information.

But technology is only as good as the people implementing it, which is why we'll also delve into the pressures teachers are experiencing in the face of generative AI's introduction to schools. And we'll take a look at how families can support educators during this transition.

By the end of this chapter, I hope you'll understand better that AI's move into schools already represents an opportunity to evaluate and support innovation, but also for families to have a say. You'll also be better equipped to examine more closely the risks and challenges that I'll outline later in this book and use that to inform your school advocacy work.

While the rapid introduction of AI into the classroom can feel overwhelming, it's actually an opportunity for all of us to help influence the impact AI innovation will have on our kids and the future. Understanding what's happening now puts us in a much better position to shape what's next, and in the end, that should feel incredibly empowering and hopeful.

Personalized Learning and AI Tutoring

If you've ever watched your child struggle with homework while thinking, "There must be a better way," you're not alone. Most of us have felt as angst-ridden as our children when seeing them wrestle with math problems or when shutting down during reading time. The pressure on families to have all the answers, especially after a long day ourselves, creates an unpleasant dynamic that shouldn't have to just exist because it's "always been that way."

It's also clear that the traditional one-size-fits-all approach to education often doesn't help the situation. With roughly thirty students in a classroom all receiving the same lessons at the same pace, using the same teaching methods, regardless of their individual learning styles, prior knowledge, or interests, some kids inevitably get left out.

Researchers for years, and well before generative AI burst onto the scene, have looked into the issues of our current singular approach to learning and have explored what it could mean to offer something more personalized.[3] There's no easy answer, which is precisely what makes AI's potential for individual instruction so compelling. Because while teachers are frequently heroic in their efforts to meet the individual educational needs of students, it's understandably not sustainable when you're managing a classroom full of kids with such a wide range of requirements.

So the idea that AI might change this is exciting—but only if we think it through and do it right. The good news is that research and advocacy for personalized learning have been swirling around for decades. So with the work having been done on why this represents a good strategy, now we need to look at how AI can facilitate this type of approach.

Celebrating "Mastery" in Education

The late American educational psychologist Benjamin Bloom first advocated for "mastery learning" in 1968.[4] His research indicated that children who were taught to individually "master" a subject achieved far more success than when learning together with their peers all at once. In his essay "The 2 Sigma Problem," Bloom suggested that 1:1 tutoring could improve children's performance by two "sigmas" in statistical terms—a dramatic leap forward.

But of course, the type of customized individual approach advocated by Bloom isn't feasible for most families. To achieve individual "mastery" of a subject would require the type of excellent (and expensive) tutors that are not accessible to most students' families. And so this is what excites many about AI's potential in education right now.

The opportunity also goes beyond the issue of access to tutors, because even the best human 1:1 instructor can only adjust their teaching approach based on what they can observe. AI systems, on the other hand, can continuously analyze dozens of data points and assess how each child might learn best. For instance, whereas a traditional math test might show that a child struggles with fractions, an AI system could reveal that it's specifically word problems involving fractions that are an issue. And even more specifically, when presented with concepts through visual representations, the student can more easily grasp the work.

Power of Content That Adapts

It's this ability for an AI system to continually "adjust" that makes the intervention incredibly effective. And we're not just talking about analyzing a student's patterns of learning but the ability to continually adjust difficulty in real-time. This adaptive capability extends beyond academic difficulty to include presentation methods too. AI can automatically switch between visual, auditory, and hands-on learning approaches based on what works best for each child. The personalization can also adapt to cultural and contextual factors that influence learning, incorporating examples and scenarios that reflect a student's background and interests.

These learning profiles develop continuously as children interact with AI educational tools. The system tracks not just right and wrong answers but also response patterns, hesitation points, preferred explanation types, and optimal challenge levels. An AI system might discover that a child learns vocabulary best in the morning but prefers audio reinforcement in the afternoon, allowing the system to adjust accordingly. It's in constant motion, which better reflects how children actually learn and grow.

Perhaps most powerfully, AI can identify learning preferences that children themselves might not recognize. That child who insists they "hate reading" might actually struggle with traditional text presentations while thriving with interactive multimedia reading content. AI systems can detect these patterns and adapt instruction to match each child's optimal learning conditions, which can considerably boost a child's confidence. It's now quite possible that Bloom's theory may finally be realized—and for all kids, not just those with the resources, time, and support.

Three Cheers for Homework Support

For many families, the idea that AI can offer homework help is a massive relief. As much as any of us support the "idea" of homework, it can be a drag to end up as academic enforcers, battling over assignments and feeling frustrated at our own inability to understand the work.

Meanwhile, professional tutoring can be prohibitively expensive and scheduling-intensive—if even available at all. Interestingly, research has found that human tutoring, even when a family has the resources to engage this type of support, can also be hit-or-miss in quality and effectiveness. According to research, while the global private tutoring market was estimated to be valued at $97.11 billion in 2023, only 15% of students receive tutoring in the US, and of those, only 2% received what would be considered "high-quality tutoring" support.[5]

There is clearly a need for a better solution. For families, the idea of a patient, knowledgeable, and tireless AI assistant available whenever children require support is undoubtedly compelling and something worth exploring.

Taking a Partnership Approach to AI in Education

While the benefits of additional tutoring are clear, research also supports the notion that this support is best when approached in the context of a larger ecosystem that keeps teachers at its core. Recent research published in Nature Scientific Reports supports this assertion, finding that AI tutoring outperformed classroom learning but that human instruction remained vital.[6] In fact, one approach the researchers explored was to "flip" the use of AI to the beginning of the learning process—using AI to teach introductory material before class. Then precious classroom time could focus on higher-order skills like advanced problem-solving and collaborative work.

We need to remember that the more emotional and sociological skills our kids need will always come from human instructors. It's the human part of learning that celebrates progress, discusses how skills connect to real-world applications, and provides the emotional support that builds lasting motivation. This is an opportunity to supplement this work, not replace it.

Opportunity for AI to Support Diversity

Often, a lack of mastery in a field is a much more complicated issue than a child simply struggling with the content. A lack of "mastery" can instead be systemic, where different groups feel excluded from a field, have different needs when approaching the content, or are made to feel inadequate overall. Nowhere do we see this more clearly than with science, technology, engineering, and math (STEM) subjects.

STEM study has been important for many reasons, but now these subjects represent a critical path to securing AI-related jobs as well. The US Bureau of Labor Statistics projects STEM jobs to grow 10.4% from 2023 to 2033—nearly three times faster than non-STEM jobs.[7] But despite this opportunity and the way these subjects have been celebrated for their importance over the years, women, in particular, still aren't making progress in STEM-related fields. According to the US National Science Foundation, women earned only 24% of bachelor's degrees in engineering in 2021, despite outpacing men in overall college completion.[8]

But many researchers now believe AI can make STEM learning more accessible, in part by providing a safer space to learn without judgment. For instance, researchers at the

University of California, Irvine, found that AI may help over-come the psychological barriers that underrepresented groups may experience in pursuing fields that are heavy on collabora-tion and multidisciplinary teamwork.[9] If kids can get past a sense of discouragement that can come from working with a group where they feel different or excluded, then they may more readily pursue a formerly-out-of-reach field of study.

Ensuring That AI Tutoring is Effective

Where to start in considering AI support for your child? First, it's important to understand that not all AI tutoring platforms are created equally. Finding the right platform can make the difference between genuine learning support for a child and expensive digital distraction that feels like an ineffective waste of time.

Look for platforms that provide detailed explanations rather than just correct answers. Quality AI tutors guide students through problem-solving processes, helping them understand the reasoning behind solutions rather than simply providing shortcuts. Also prioritize solid safety and privacy features—as we'll cover in Part Two, privacy, safety, and trust are crucial factors when choosing any platform.

Once you've found an AI system you trust, successful inte-gration requires thoughtful planning about when and how your child should access AI assistance. As you get comfortable with this approach, the next logical step is evaluating how effective the AI intervention has been. Evaluation of student learning is another area where AI sits on both sides of the equation—we must consider AI's effectiveness as a teaching tool, but AI tech-nologies also power these evaluations across all types of instruction.

How to Decipher AI-Powered Learning Insights

Remember when a report card with letter grades felt like the complete picture of how your child was doing in school? Those days are quickly becoming a thing of the past. If you've recently received a progress report that looks more like fitness tracker output than a traditional report card, or if your child's teacher has started talking about "learning analytics," you might be witnessing the shift toward AI-enhanced assessments.

This change can feel confusing, and some families are rightfully suspicious of the accuracy and true efficacy of these types of assessments. But when you think about the possibilities, it does make sense. For instance, now instead of a simple "B+ in math," you might learn how your child approaches problems, how fast they grasp a subject, and other critical engagement patterns that put a grade into context.

The efficiencies of automated and personalized feedback can work both ways as well, providing critical feedback to educators. Research published in *Educational Evaluation and Policy Analysis* in 2023 found that AI-powered tools for evaluating teachers themselves both supported instructor professional development and greatly increased student satisfaction and performance at the same time.[10]

Evolution of Educational Assessment

Traditional assessment to date has been like sitting down for just one annual (or even quarterly) family portrait. You got a clear picture at that specific moment but then missed everything that happened in between. Think of AI-enhanced assessment as more like having a video camera running, capturing learning as it happens in real time. This might mean receiving more frequent updates about your child's progress or even getting weekly or daily insights into how a child is doing with specific skills or concepts.

You might be thinking, "Isn't this information overload?" And yes, it can be if you are trying to analyze every single data point. But the key here is the ability to focus on patterns and trends rather than getting caught up in daily fluctuations. AI assessment systems can also track things that traditional tests cannot measure. They might show how long a child typically spends on different types of problems, which concepts they've mastered versus which ones they're still developing, or what time of day they tend to be most focused and productive.

This detailed information can help answer questions that letter grades cannot. For instance, why a child might have struggled with a particular math unit or whether they actually understood the material or just memorized the steps. It's all about a better grasp of the learning process in a more efficient and actionable manner. It should serve as a supplement, not a replacement for a family's understanding of their child as a learner.

Rethinking Performance Metrics Today

AI learning assessments can also be a game-changer when it comes to family discussions about academic performance overall. Imagine that instead of asking, "How was school today?" and getting a shrug, you could leverage these detailed assessments to ask more productive questions about the work your child did during a school day.

Many families feel trepidation about turning their children's learning into data points to be analyzed. But we have always analyzed kids' performance, just not very efficiently or specifically. Now it can be done with more precision and actionable output. Remember, kids often struggle to know why they are having trouble learning a subject, and this can cause a lot of tension and unrest at home. The hope is that continuous monitoring (and whatever that means to a family) catches learning difficulties early and before they become major problems.

Traditional assessment might also miss big issues like reading struggles or undiagnosed learning challenges until annual testing takes place. Here AI systems can detect early warning signs like declining comprehension scores or increasing time spent on vocabulary recognition, for instance. This early identification allows for immediate intervention, which can be as welcome for a student and their family.

Data Privacy Concerns of Assessment Tools

With all this detailed tracking come legitimate privacy concerns. Schools are collecting an unprecedented amount of data about how kids think, learn, and even behave (and starting well before this most recent AI era). It's important to understand what information is being gathered, how long it will be stored, and who has access to this information.

School-based AI assessment platforms are subject to student privacy laws like the Family Educational Rights and Privacy Act (FERPA), which we'll discuss in later chapters. But the challenge, as I've mentioned earlier, is that many of these regulations were written before AI-enhanced learning became more widespread. Frankly, even before the Internet burst onto the scene. And this is precisely why families should stay active and informed about the tools being used, their privacy policies, and how schools interpret and enforce existing laws and regulations.

Keeping up With Rapid Change in AI Systems

If and when your school starts to use more sophisticated AI-powered technology to assess kids' progress, you shouldn't hesitate to jump in with questions. Teachers are learning to interpret newer AI data analysis tools just like we are and frequently welcome the increased engagement by families.

It's also important for families to feel comfortable with what is being assessed overall. You might feel that math is okay, but any type of social and emotional testing is not something you are comfortable with your children participating in. Not only do families have every right to ask the hard questions related to these tools and surveys, or make the request to opt out, frequently this can result in schools reconsidering these tools overall.

While detailed progress data can be incredibly helpful, remember that it's just one piece of your child's educational story. Their creativity, collaboration skills, curiosity, and character development might not show up in learning analytics, but they're equally indispensable to a child's overall growth and success. Embracing AI for these purposes does not mean overlooking progress in softer skills for lack of an ability to measure these areas.

The goal of AI-enhanced assessment isn't to turn children into collections of data points but instead rich information to help better understand how children learn and what support they need to succeed.

These new assessment approaches work best when combined with a more in-depth understanding of a child as a whole person, not just a student. But probably more importantly right now, we need to keep our fingers on the pulse of how technology is used in this manner. There is a fine line between assessment usefulness and invasive analysis and this is best determined by individual families working with the educators who know their kids best.

How to Support Educators in Tackling AI

Understanding how teachers experience this AI movement is crucial for families who want to advocate for balanced policies. Teachers have found themselves at the center of an AI maelstrom, and many are struggling to catch up or feel heard about what they see as useful to the job they do. Teachers are also often expected to implement policies they didn't create, manage technology they may not fully understand, and navigate student AI use without adequate training or support. For some teachers, this feels like a tipping point—and not in a good way.

The reality is that educators are caught in an impossible position right now and without the opportunity to even consider the upside of AI. Instead, they are expected to become immediate experts while simultaneously excelling at their core responsibilities: teaching, managing classroom behavior, supporting diverse learning needs, and navigating increasingly complex administrative demands.

According to an *Education Week* Research Center survey from December 2024, 60% of educators said their districts still hadn't made their AI policies clear. An equal number felt that their students didn't have the necessary direction here either.[11] Imagine trying to do your job effectively when you don't even know the rules and boundaries—it would be difficult to get excited about anything in that type of environment.

At the same time, though, AI offers enormous potential that could impact teacher job satisfaction and the quality of their instruction in numerous ways. And most notably (but less discussed) in reducing the burden of administrative duties that can keep educators away from the job they signed up to do—teach our kids. In fact, according to additional research from *Education Week*, teachers spend up to 29 hours a week doing non-teaching, administrative tasks rather than in direct instruction or student interaction.[12] In supporting teachers to explore and decide what they might find most useful about AI systems to the important work they are doing, we can ensure a win-win for everyone.

Shape-Shifting AI School Policies

It's fair to say that right now the primary focus related to AI school policy is on generative AI tools in the classroom (and at home for homework). But by focusing so narrowly, this may overlook a far bigger picture related to AI and other technological innovation coming to the classroom. It's a sentiment experts share too. For instance, the Digital Education Council, a consortium of more than 100 global higher education institutions, noted in its most recent report that by focusing only on generative AI, schools are taking too limited of an approach.[13] The risk, of course, is in missing the much larger array of AI-powered tools that are already operating in school environments and a rapidly shifting technological landscape that will bring new advancements too.

Most school AI policies also often fall into predictable patterns that reveal their underlying approach to AI integration and can be insufficient or cause even greater confusion. For instance, a blanket ban on generative AI or a more vague permission model that only allows for AI system use "with teacher permission" or "when appropriate." Or policies where kids are required to disclose AI use (which gets confusing when you consider tools such as Grammarly are technically AI-powered).

Problem of the AI "Cheating" Assumption

Current school AI policies can also embed assumptions about learning, technology, and student capability that may not align with a family's educational values or evidence-based practices. For instance, the assumption that AI is "cheating." This one, of course, is particularly problematic as it may mean schools employ "AI detectors" to evaluate students' work—which can be inaccurate and cause kids immense distress.

According to Northern Illinois University's Center for Innovative Teaching and Learning, most AI detection platforms report a false positive rate of 2-3%.[14] Even at such a small percentage rate, these inaccurate results can quickly add up and impact hundreds of students during a given school year. These tools have also shown bias against non-native English writers, neurodivergent students, and even students who might insert regional dialects into their text.

This cheating assumption can also result in immense stress for kids who fear being falsely accused of using AI inappropriately and for teachers who don't want to engage in this type of conflict either. It can also have the unintended consequence of stifling the development of a student's "personal voice," encouraging them to focus too much on how to evade AI detectors, even when they didn't use AI. It's an unfortunate situation that isn't serving anyone, and it's something we'll discuss in more detail later in this book.

How Teachers Experience Technology Policy

The hardest part for many teachers is being asked to be experts in a domain that even researchers are still figuring out. Most teachers also report that AI policies are developed at an administrative level without their meaningful input, creating implementation challenges that families rarely see. They're expected to enforce policies they may not even fully understand while managing student AI use they cannot always detect.

This disconnect is creating significant trust issues. According to a 2024 Center for Democracy & Technology survey, 52% of teachers said generative AI has made them more distrustful of whether students' work is authentic—and that number jumps to 69% when schools have banned the technology.[15] The lack of consistent policy is also contributing to an environment of stress and mistrust that ultimately affects student learning.

Consider the position this puts teachers in: they must evaluate student work for potential AI use without a clear understanding of how AI systems function, what AI-generated content looks like, or how to distinguish between helpful AI assistance and problematic dependency. It's like asking someone to referee a game without explaining the rules.

Addressing Teachers' Job Uncertainty

Teachers' concerns about job security aren't unfounded. New schools are opening with the explicit promise of revolutionizing education through AI-powered personalized learning, with little to no direct teaching staff. When technology leaders speak enthusiastically about AI tutors that "never get tired, never lose patience," the implicit message is that human teachers are inherently flawed and replaceable.

These additional pressures are landing on a profession that was already struggling with the weight of expectations juxtaposed against tight salaries. According to the National Education Association's 2024 report, the average teacher salary is $72,030, and in most states, the pay hasn't kept pace with inflation.[16] Now educators are expected to become technology experts, data analysts, and AI literacy instructors, all without additional compensation?

Families often don't help relieve the pressure teachers face either. We frequently demand transparency about AI tool use but then simultaneously expect teachers to prepare children for an AI-integrated future. Some of us advocate for embracing AI tools, while others insist on protecting children from potential AI harms. Teachers bear the brunt of any frustration families have over changes they had no control over implementing in the first place.

What Educators Want and Need to Manage AI

Contrary to the assumptions that may come from stories about teachers bemoaning "AI cheating," educators are enthusiastic about AI innovation. Their concerns stem from a lack of support and guidance rather than opposition to innovation. According to 2025 research by Carnegie Learning, teacher optimism for AI has grown, with 81% somewhat or extremely optimistic about AI in education, up from 67% in 2024.[17] But the training opportunities are still not there, with the report finding that 57% of school districts nationwide still hadn't provided sufficient AI training to teachers.

What educators require is clear, specific, actionable guidance. They want detailed policies rather than vague statements that leave interpretation to individual judgment. Teachers want to know exactly what AI uses are encouraged, acceptable, or prohibited, with specific examples they can apply consistently across their classrooms. They also require training on AI-specific educational applications. And perhaps most importantly, teachers must have input into policy development rather than just comply with top-down mandates that ignore classroom realities.

When teachers have the support, training, and voice they require, they can become powerful advocates for thoughtful AI integration that serves students well. The key is giving them the tools and agency to make it happen.

Hope for Administrative Support and Efficiency

Arguably the area where teachers may experience the most support and welcome the help AI provides is on the administrative side of their jobs. We often talk less about this because it's not the tools and technology that "touch" our kids, but it's arguably just as important. When educators can reclaim hours of their time spent organizing lessons, grading, or other administratively heavy work, they can put more time back into classroom work.

AI can dramatically reduce teachers' burden in a number of ways. It can automate lesson plan creation, generate quiz questions aligned with curriculum standards, and provide detailed feedback on student assignments. For grading specifically, AI can handle the initial review of essays, math problems, and even creative projects, flagging areas that need human attention. This doesn't replace teacher judgment but can enhance it, allowing educators to focus their expertise on nuanced feedback, relationship building, and the creative side of instruction design.

How Families Can Support Educators Today

As I'll share in Part Three of the book, by understanding the pressures that teachers face, we can help support both our children's education and the educators who serve them by advocating for supportive school policies. We can also approach conversations with teachers from a place of partnership rather than judgment. When your child's teacher is trying a new AI tool, ask how you can support the effort at home. When policies seem unclear, advocate at the district level for better guidance rather than putting pressure on individual educators.

Rather than seeing teachers as obstacles to overcome, we're better off approaching them as partners working toward the same goal of effective education and student success. Most importantly, when we support teachers, we support our children. And right now, educators need our advocacy more than ever. There are remarkable opportunities ahead to make teachers more effective, especially in creating a more productive, optimistic, and inclusive classroom.

Exciting Ways AI Can Foster Classroom Inclusivity

Many families today are in the privileged position of looking at AI-assisted learning as a "nice to have" because their children already benefit from a solid education or don't require any specific interventions. But for families with a student who does need the extra support, AI can be truly transformative—both in the tools that foster actual inclusion but also in providing the most robust support that many schools find financially prohibitive to offer.

According to the National Center for Education Statistics,

roughly 15% (7.5 million) of all public school students in the US require special educational accommodations.[18] And of these children, more than 2.4 million have a "specific learning disability" (SLD).[19] At the same time, many school districts struggle to hire special education teachers. According to a 2024 report from the US Department of Education's National Center for Education Statistics, 74% of elementary and middle schools and 66% of high schools reported difficulty finding fully certified teachers to fill special education teaching roles.[20]

The reality is that accommodations in most schools are inconsistent, unavailable, or, again, prohibitively expensive, with many families having to frequently fight to receive the support their child requires.

Potential to Address Accommodation Challenges

My friend Serena's daughter is in middle school. She said that the process of receiving accommodations for her child's dyslexia had been exhausting and that she hoped schools would consider all children when considering the use of AI tools in the classroom. "I hear a lot of pushback related to AI in schools," she said, "but no one seems to be talking about how kids with learning challenges might benefit from the new technology available."

Many families of special needs kids share Serena's hope that AI might unlock more robust, cost-effective, and equitable support that not only helps their kids but keeps them with their peers in the classroom. The numbers highlight the optimism that these parents feel at the prospect of AI support. According to the Special Olympics Global Center for Inclusion in Education, 77% of parents and 64% of educators believe AI will make education more inclusive for children with intellectual and developmental disabilities.[21] So, as we all figure out AI's role in schools, we shouldn't forget those families that might find the extra support a game-changer and feel an increased sense of urgency to adopt these AI systems.

The Specific Innovations Transforming Inclusion

Just as AI systems can act as intelligent tutors to all kids, they can be more precise in identifying where students with special needs may require additional support. For instance, for kids with dyscalculia (difficulty with math) or dyslexia, a system can adjust its difficulty in real-time, providing alternative explanations and offering different practice problems. This level of customized instruction is often too expensive for many families and districts, where the only option is 1:1 tutoring support.

The options for addressing personal needs are endless. For children with challenges such as ADHD, this can look like complex tasks broken into smaller, manageable steps or frequent breaks and movement opportunities. For kids with autism spectrum disorders, AI can offer consistent, predictable responses and clear, structured learning pathways. AI tools also don't display confusing social cues, which allows these students to focus on learning content rather than navigating social interactions.

The way in which many more children's individual needs can be met may also give families and educators more flexibility in having a wider range of learners in a single classroom. This can come as excellent news for kids who would have otherwise been separated from their peers. A few more exciting examples:

VERBAL COMMUNICATION ASSISTANCE

With AI support, students who struggle with verbal expression or written communication can more actively participate in classroom discussions, answer teacher questions, and contribute ideas without being held back by communication barriers.

For children with dyslexia or other writing-based learning differences, AI can provide predictive text, contextual grammar support, and speech-to-text input. This means a child can express complex, higher-order thinking without being limited by spelling or handwriting challenges. Students who struggle with verbal expression can use AI-powered communication aids to help formulate and organize their thoughts before speaking.

Advanced text-to-speech systems can also simulate natural intonation with customizable features like adjustable speeds and synchronized word highlighting, helping students with reading difficulties access written content while reinforcing word recognition and comprehension.

SEEING AND HEARING THE WORLD DIFFERENTLY

AI-powered applications like Microsoft's Seeing AI and Be My Eyes can be life-changing for students with visual impairments. These apps provide real-time audio descriptions of people, surroundings, documents, and digital content, letting these children navigate classrooms, libraries, and assignments more independently. Not only can these tools allow users of the technology to better engage with the world around them—it can keep children in the classroom with their peers.

For students who are deaf or hard of hearing, tools like Otter.ai provide real-time transcription of classroom lectures, while Google Live Transcribe offers speech-to-text conversion. Live captioning integrated into video conferencing platforms like Google Meet and Microsoft Teams also now provides speaker identification and context-sensitive transcripts, helping students follow lectures and discussions more easily. By reducing reliance on interpreters, these tools give many children a level of freedom they've not known before.

BREAKING DOWN LANGUAGE BARRIERS

For English language learners, AI can provide real-time translation, pronunciation guides, and culturally sensitive explanations that help bridge the gap between home language and classroom instruction. These tools can identify when a student is struggling with language versus content, ensuring they receive appropriate support rather than being misidentified as having a learning disability.

AI can also help teachers understand cultural contexts that might affect learning, providing insights into different educational backgrounds and recommending culturally responsive teaching strategies.

Power of "Universal Design" Approach

AI can take standard classroom material and transform it in ways that adapt to individual students by presenting the same material in different ways. AI systems can automatically simplify complex text, summarize key points, or even generate interactive 3D models of concepts for students with cognitive challenges. A more visual learner might access information through dynamic diagrams and interactive visualizations that make abstract concepts concrete. And a student who learns better by listening can receive content through natural-sounding narration and conversational explanations.

The point is AI can offer kids the chance to study the same curriculum but in a way that works for their learning style. This approach is called Universal Design for Learning (UDL).[22] Think of it like designing a building with ramps, elevators, and stairs from the start. UDL recognizes that having various learners in any classroom should be the norm, not the exception. Ultimately, these systems can create one seamless experience, accommodating all learners while creating efficiencies that make the technology more accessible and cost-effective.

AI and Early Detection of Learning Challenges

AI systems aren't only excellent at adapting but also at helping to identify kids' challenges in the first place. By analyzing patterns in student performance data, such as test results, reading speeds, assignment completion rates, and even engagement levels, AI can flag early signs of dyslexia, ADHD, language-processing delays, or comprehension struggles. This can save families and children countless hours of stress, anxiety, and uncertainty and quickly get them the support they require.

This also means teachers can intervene earlier and tailor support strategies before a child falls behind. For students already receiving accommodations, AI can track which interventions work best, making data-driven recommendations to adjust supports dynamically and optimize learning outcomes. This approach replaces the old "wait-to-fail" model with a continuous feedback loop, benefiting both students and educators.

A Rising Tide Lifts All Boats

One of the greatest results of AI-driven accessibility is the possibility of normalizing accommodations. Instead of singling students out, AI tools are designed for all learners in very personal and discreet ways. And even when more sophisticated support is required, it may mean that children with special needs can actually stay in the classroom with their peers who don't require extra support.

The key is ensuring that as AI tools develop, they're designed when possible with all learners in mind from the beginning—not as an afterthought. When we embrace AI accessibility tools thoughtfully, we help ensure that every child can access learning opportunities that match their potential, regardless of their learning differences or challenges.

And that's one thing that I hope gets remembered in debates around AI's use in the classroom, because it can mean so much more than tech-enhanced learning; it can mean kids learning together no matter the challenge or gift. And by advocating for inclusive design and pushing for comprehensive AI accessibility features, we can create learning environments that work for everyone.

Key Chapter Takeaways

If you're feeling a bit overwhelmed by everything we've covered in this chapter it's more than understandable—it reflects how many of us feel each day in considering all of these new tools and issues. From the potential of personalized tutoring that adapts to a child's learning style and the possibility of accommodations that happen in real-time, combined with confusion over many schools' actual AI policies, it's a lot.

The newness of the opportunity and the speed with which it's been evolving is what's most disruptive. Ironically, the technology itself is straightforward to understand, and, as we'll talk about in coming chapters, the work we need to do is to make sure AI is used right and with privacy and safety in mind.

What's most important to acknowledge in the educational realm right now is how difficult it is for all of us to provide confident, informed feedback on AI technologies in schools when we're just catching up ourselves. There is also a tension between the adults approaching AI with caution and the young early adopters in schools diving right in, becoming proficient at warp speed. As my friend Damien said, it's almost impossible to have an opinion about issues such as AI and academic integrity when we don't have the basics squared away ourselves.

But I hope that you are now better equipped to engage in conversations about AI with educators, other families at school, and your kids. Starting to understand the landscape should help us approach the challenges and opportunities ahead with a balanced perspective. So let's recap some of what we covered specifically in this final chapter of Part One.

AI as a Partner Not a Replacement for Teachers

The most effective educational AI applications work as collaborative partners rather than replacement teachers. AI excels at providing detailed explanations, endless practice opportunities, and immediate feedback, while human teachers and families offer motivation, wisdom, emotional support, and the kind of complex guidance that helps it all come together.

This partnership approach recognizes that education is fundamentally about human connection, enhanced by technology rather than replaced by it. When AI handles routine tasks like drilling math facts or providing reading comprehension practice, teachers can focus on higher-level instruction, creative projects, and building stronger relationships with students. The same is true with homework support at home, with AI helping to relieve the stress of providing tactical support and letting families focus on the big picture.

AI's assessment capabilities can also provide families and educators with insights into a child's learning patterns, progress, and needs, unlike anything previously available. Rather than waiting for periodic report cards or standardized test results, you can get real-time information about your child's academic development, learning style, and areas where they require extra support. Such awareness can transform conversations about education from general discussions about grades to something much more actionable and productive.

Chance to Keep All Students in the Classroom

AI's accessibility features are especially exciting for the ways they may level the educational playing field like never before. Children with learning differences, diverse cultural backgrounds, varied learning styles, and different economic circumstances now have the potential to access high-quality, personalized educational support that can adapt to their specific needs. And it can do this naturally and invisibly, removing barriers without making a child feel different or singled out.

The economic implications are equally significant. For most American families, private tutoring or specialized learning programs are simply not an option. But AI-powered educational support can provide many of these same benefits at a much reduced cost. While AI can't eliminate educational inequities and we need to watch for bias, as we covered in Chapter One, AI represents a powerful opportunity for expanding access in previously unthought-of ways.

Navigating the Educational Challenges Ahead

While AI's educational potential is notable, implementation is not without its challenges. Schools require thoughtful policies that balance innovation with privacy protections. Teachers need robust professional development to effectively integrate

AI tools into their instruction. And as families, we need to be patient, engaged, supportive, and open-minded to the potential.

By understanding the opportunities and responsibilities that come with these powerful tools, we can ensure that AI's benefits don't come at the expense of human interaction and in-person instruction. We can also better serve as advocates and cheerleaders for our kids and the teachers who are critical to their educational development.

The balance required here is seeing the opportunity but also not losing perspective on the challenges and risks inherent in these technologies at the same time. In Part Two, we'll explore in more detail some of these issues and the downsides of AI if important consequences are not considered, such as an invasion of data privacy or the impact on our kids' mental health.

Part Two

Knowledge is Power: Navigating AI's Challenges and Risks

Chapter 4
How AI Can Impact Your Family's Privacy

Recently, my friend told me about a moment that thoroughly changed her understanding of how interconnected our digital lives really are. Her 15-year-old daughter had been struggling with a mental health issue, and together they had found an online platform that was providing real support. But a few weeks later, something unsettling happened.

While her daughter was watching a movie with friends via a streaming service, targeted ads began appearing for medications and mental health services related to her exact condition. The ads were specific enough that her daughter immediately recognized the connection and was mortified. What made this particularly painful was that she had been intentional about keeping her struggle private. Seeing those ads felt like a violation—as if her most personal struggles were being broadcast back to her during what should have been relaxing downtime with friends.

My friend was shocked to discover that her daughter's private health activity online had influenced the advertising she saw on an entirely different platform. "I always thought these services kept information separate," she told me. "I didn't realize that seeking mental health support could follow my daughter around the internet." The experience led them to cancel the streaming service entirely and read all privacy policies much more closely.

My friend's experience reveals just one layer of how AI-powered systems gather information about our families. And when we talk about AI in this capacity, we're not always just talking about generative AI chatbots but a vast ecosystem of data collection powering both new and old platforms equally. It's AI's effect on the "behind the scenes" part of our digital experiences that can come as a surprise to many families.

This also represents a fundamental shift from earlier technologies, which largely stored and used only the information we explicitly agreed to provide. Now our digital activities become the "fuel" that powers AI systems, often through confusing terms and conditions and without our full understanding about how information travels between platforms—or the unintended consequences that follow.

Data Implications of the Rush to Build AI

The goal of data collection is generally to create products and services that are more useful, efficient, and personalized. Personal data can enable tailored learning opportunities, accelerate medical discoveries, and provide extra support for those who need it. But achieving this potential means seeking more and more data—and typically from anywhere it can be found and sometimes without our knowledge that it is happening.

A large portion of what's used to "train" AI was collected during a different era of digital consent too. Most of us didn't fully understand what we were agreeing to when we shared data online 5, 10, or 15 years ago. Frankly, the companies collecting that data didn't always make it clear either.

What's more, high-quality data is becoming increasingly scarce, with less than half of the data available on the internet considered "good enough" to use for AI system training purposes—something we discussed in Chapter One. Researchers also now expect that the overall supply of web-based data will be depleted sometime in the next seven years, and potentially as early as 2026.[1] This, of course, creates additional pressure on consumers in an already complex privacy environment online. But in understanding how AI collects and uses family data, we can help shift the balance—and understand when more data than we want is being taken from us.

Managing Our New Data Privacy Reality

For many families, the realization of how much data AI systems collect can feel overwhelming. But the truth is, when we use a digital service, we always "pay" for it—in the form of our data, as money, or both. This has been the case for many years. It's just that now, in our new AI-led reality, the consequences of not properly considering these data decisions are far more serious.

Here's what matters though—each one of us controls this data exchange, even if we haven't paid as close attention as we should. You can decide what personal information you're willing to hand over in exchange for a service or tool that benefits your family. We are the consumers and need to start acting more intentionally when we're online.

The challenge isn't just about individual privacy either but an entire family data ecosystem where all of our information creates insights about one another. When AI systems know that your teenager is a new driver or your spouse has mental health challenges, they can build profiles with real-world consequences. In January 2025, for example, Texas Attorney General Ken Paxton sued Allstate "for unlawfully collecting, using, and selling data," such as location data, and then using this information to raise insurance rates.[2] It's not the first time this has happened, and it won't be the last. As we continue to use services without checking the associated fine print, these practices will only become more prevalent.

What You Will Discover in This Chapter

The challenge for many families lies in recognizing that data collection in the AI era extends far beyond obvious activities like filling out forms or creating new accounts. AI systems gather information from our search histories, the time we spend reading articles, the photos we take, the apps we download, and even, in some cases, how we move our mouse cursor across a screen. But once we know what is collected, how it's collected, and why—we gain a new freedom in the power to actually choose what we do next.

In this chapter, I want to show you what having this kind of choice really means. You'll see concrete examples of data collection that families often don't recognize, like how voice assistants learn from what they hear or why free educational apps and email services often collect far more information than the paid alternatives.

I'll walk you through how digital consent actually works in practice, moving beyond those endless terms of service agreements we all scroll through without reading. You'll understand in more detail what families really agree to when clicking "accept" and how to recognize when you are the customer versus "the product"—a distinction that matters now more than ever.

You'll also discover why children's data has become even more valuable to this new digital economy and why kids' information can be so difficult to delete or control once collected. By understanding all of these dynamics, families can make more informed decisions about which platforms and services align with our values and meet our future needs.

By the end of this chapter, I hope you'll feel more confident approaching your digital life, even beyond the new AI platforms we are focusing on today. You'll understand not just what's at stake, but what choices you have—and how to make decisions that feel right for your family. Let's start with what I think might surprise you the most: the scope and scale of specific personal data being collected about us right now.

What Personal Data AI Collects

You might think you have a pretty good handle on what family data is collected online. I thought I did. We consciously share information when we fill out forms, create profiles, or upload photos to social media platforms. We understand targeted ads and streaming recommendations. The most diligent of us delete unused accounts and keep our passwords strong. But generative AI has changed the equation entirely, and it's not always that apparent what is occurring.

These systems need enormous amounts of data to be effective, and AI-powered collection methods make gathering the information easier and far less visible than ever before. The result is a cycle where AI creates the need for more data while simultaneously making it simpler to collect—it's literally the reason that more of our personal data is in demand and the mechanism by which it's more easily scooped up. Dizzying, right?

Here's the thing though—we've been around this digital neighborhood for quite some time. AI innovation is just forcing us to level up our awareness and put better protections in place. And this isn't about creating fear or rejecting the digital products and services we want to use. It's about shifting the power dynamic and giving more of it back to us. We need to stop thinking of ourselves as passive consumers but instead as powerful brokers of our data.

Every digital interaction is now an opportunity to decide if the trade of our data is worth it for what we get in return. It's possible to change the game, too, especially if we do it together. We just need a far clearer picture of what's actually happening.

What AI Wants from Us

To understand why data collection has intensified so dramatically, it helps to think about what AI systems actually need to function. If you were tasked with building a robot, what information from a human would be useful? It's a question I asked in an article nearly a decade ago.[3] And the answer is everything —a rich and varied collection of data that reflects the full picture of human behavior. When you think of AI in the broadest terms as a technology that seeks to replicate the best of human beings, then it should be easy to imagine what data is necessary to create technology that successfully "mimics" us. From how we think to how we act to who we are at our core, AI systems want it all. And technology builders will jump over plenty of hurdles to get this information.

Today's AI technology analyzes the patterns in how we communicate, learn, and behave. It tracks the hesitation in a child's typing, the confidence level reflected in word choices, and the emotional undertones in how questions are phrased. When your child writes, "I think this is wrong, but..." AI systems can note not just the content but the self-doubt embedded in that language.

Reading Between the Digital Lines

These systems don't stop at what we explicitly tell them. They build inferences about our cognitive patterns, emotional states, and social dynamics. They note references to family structure, economic context clues, and developmental changes over time. The data includes not just what children know academically, but how they learn, where they struggle, and what motivates kids to persist or give up.

AI systems do this by collecting information about when we're online, how long we spend on tasks, and patterns in our digital habits—what researchers call "behavioral metadata." This information can reveal sleep patterns from device usage times, stress levels from how quickly we respond to messages, and social confidence from whether we seek help immediately or try to solve problems alone first. For children specifically, this data collection can reveal things parents might not even recognize about their kids' personalities, learning styles, and emotional development.

This data isn't collected to be used against us but to make a digital tool or service better and more effective. The challenge is that these insights might extend far beyond what we originally opted into sharing or understood that we were sharing. And if we don't start paying attention, the dynamic will simply carry on until we rein it in.

Old Data That's (Unexpectedly) New Again

AI systems collect data in familiar ways, but they also gather information continuously through everyday activities we might not even think about. They can even work with data that was shared years ago and information that seems insignificant. Former users of the photo-sharing platform Flickr discovered this back in 2019 when news broke that IBM had used more than a million personal images to train its AI systems.[4] When users originally signed up for the service, they had unknowingly signed over control of their images. Those vacation photos and family pictures that seemed harmless at the time later became part of training content for computer vision technology.

The lesson here is that there's really no such thing as "useless" data anymore. Comments left on a friend's social media post ten years ago, your teen's post on Reddit when they were in middle school, or even metadata from 2004 photos you long forgot can be valuable for training AI. What seemed trivial or temporary when we first shared it can now be repurposed in ways we never anticipated.

This means it might be time to rethink what you've already put out there, not just what you're sharing today. All of these past digital breadcrumbs create detailed profiles of our families that go far beyond what we ever intended to share. And at the very least someone is profiting from this detail, and it's unlikely to be you.

How Chatbots Collect Data

Now we aren't just dealing with an always-on data ecosystem. We're also navigating the unique risks that come with conversational AI. The human-like "voice" of chatbots creates an illusion of casualness and encourages friendly exchanges. Behind the scenes, algorithms may be analyzing response times, hesitation patterns, and emotional cues in word choice and phrasing.

AI chatbots can be built responsibly with minimal data collection. But it's builder-dependent, which means there undoubtedly will be platforms operating without the necessary guardrails in place. And without preparing kids properly, they are vulnerable to the "human-like" tone of chatbots. In fact, researchers have found that children using highly personalized chatbots that seem friendly and responsive are indeed more likely to share too much personal information because the interaction feels so safe.[5] The parallel to teaching kids about talking to strangers is obvious, but now the "stranger" sounds friendly, helpful, and safe.

Reality of What We (Often Unknowingly) Share

Most families underestimate the current sophistication of data collection. We've lived with privacy tradeoffs in technology for years, but now, as more data is required from us, things are getting more intense without our notice.

We've often convinced ourselves that there's "nothing special" about our information or that we "have nothing to hide." We also put too much faith in the platforms and brands we've come to trust. But if you take some time to check out the terms and conditions you've opted into for any platform you use, the implications start to become crystal clear.

Just consider, for instance, that all the photos you have in a public Instagram account are available for Meta to use in training their AI systems.[6] While users do "own" their personal photos and can delete them, once images are incorporated into training data, their contribution to the AI system can't be undone—even if you later remove the content from your account.

Even worse, while users in some countries have been able to opt out of having personal photos used for AI training purposes due to stronger privacy regulations, US families have fewer protections. But families can make their accounts private, and that's an excellent example of how one simple action can give us back more control of our information.

There are many more examples of how data collected for AI training purposes technically sits within terms and conditions we previously agreed to. The problem is that the language has been so purposely broad and vague as to allow for shifts in how the data is used. So it might be time to rethink how quickly we opt in going forward and take a closer look at the terms and conditions of the platforms we use today.

The fact is, the data that powers AI's personalization is also what makes these systems so helpful and engaging. So the question becomes: how do we balance our desire for AI's benefits with our comfort level around data sharing? Understanding what gets collected is just the first step—and a good one. But we also need to better understand how data is gathered in practice, because that's where the real opportunities for protection live. And that's precisely what we'll tackle next.

How Data is Collected by AI Systems

While it's good to know what AI systems collect and to be certain we consent to this collection (something we'll talk more about later in this chapter), the real starting point is in understanding how it happens overall. This is the key to everything, from participating more confidently in conversations about AI policy at school to making choices that truly align with our family's values and beliefs. And it's not complicated either.

Think of it like understanding how your home's electrical system works. You don't need to be an electrician to know which breakers control which rooms or to make smart decisions about what appliances to plug in where—or even how much we should expect to spend on energy. The same principles apply here. You definitely don't need to become a data scientist to make informed choices about your family's digital life. And once you understand how everything works, you'll start to recognize those data collection practices that ultimately don't work for your family.

Data Collection Methods

Data collection methods keep changing, but the fundamentals stay consistent. Once you understand how these basic methods work, you'll be able to spot new collection tactics as they show up.

DATA "MINED" THROUGH AI CONVERSATIONS

Every AI conversation is a data collection opportunity for a company trying to improve their product and ensure their services work for you. To understand this is not to reject the experience but to approach it knowingly. Unfortunately, children often don't recognize what's happening behind the scenes when chatting with AI systems (as we just explored), but really many adults don't either.

Because chatbots are so new and the conversational tone feels natural and encouraging, it becomes far too easy to share personal information freely with these systems. The interaction feels safe and informal, and many of us haven't developed the instinct to stop and consider the bigger picture. That said, personalization, as we've discussed, is the key to AI systems' effectiveness. So just like anything, the more you know, the better you can understand the tradeoffs.

For example, when a child tells an AI tutor that they're stressed about an upcoming test, the system may respond with calm encouragement while also noting details like vocabulary level, emotional tenor, and the timing of when help is sought. Those may be useful data points for an AI tutor that a family might be okay with.

Similarly, generative AI functionality embedded in social media platforms can also "mine data" from conversations online, from comments and posts to engagement patterns. And now increasingly, with AI chatbots in the mix too. This type of data mining for personalization may not work for your family. And it's important to know the difference.

When we don't understand what's happening during these exchanges, it's impossible to make the kind of informed and empowered decisions we need to about what we are sharing. And not all AI-powered applications are equal.

Inference Gained Through Usage Patterns

AI systems don't just collect what users tell them directly either. They also "infer" information from how people behave. Think of it like a teacher who notices patterns—not just from what a student says, but from when they raise their hand, how often they ask for help, or whether they work alone or in a group.

Every pattern tells a story—when your child logs on, how topics shift during a session, which phrases they repeat. These patterns can also reveal family routines, whether parents help with homework, or even a child's emotional state based on consistency and tone.

The tricky part is that these inferences happen invisibly. Your child isn't choosing to share this information. The system is piecing it together from behavior, building a profile that extends beyond the homework question that brought them there in the first place.

You might find the very idea of such inferences being made about your family problematic, and that's completely understandable. But, again, this type of pattern analysis is happening across many of the platforms you're already using, including social media and entertainment streaming services. A clear understanding of both the mechanics of inference and how many companies employ these tactics helps ensure a consistent approach.

As with overall data mining, these "signals" can help AI systems support you in the ways you need most. But again, it's about making a conscious choice and considering what you might be getting in return.

BIOMETRIC AND BEHAVIORAL ANALYSIS

Adding complexity to this already deeply intricate ecosystem is the ability to collect new data that's like nothing we've come across before. In this case: biometric and behavioral analysis. This type of data is often overlooked but one of the most critical categories of sensitive information for families to protect.

The fact that the AI systems built into everyday platforms like social media can also perform this new type of data collection is unnerving. We are talking here about the tracking of voice tone, maybe typing speed, facial imagery, and even our expressions through device cameras that can infer mood or even mental or physical health indicators.

Even more unsettling is the ability for new technology to identify new data points such as "micro-expressions." These are facial ticks or slight movements that denote feeling without a user even realizing it.[7] The implications are quite breathtaking. From the ability to tell if someone is lying to finding that those cameras at checkout might ultimately be there to identify how you feel about each item that you are purchasing. As a marketer, you might love the idea of potentially using this biometric information to, say, extend customized offers to shoppers. But the problem, of course, is that such a mismatch between users' expectations and this type of commercial reality does not make for a transparent and harmonious relationship.

SURPRISING EXAMPLE IN SOCIAL MEDIA BIOMETRIC USE

Social media platforms have used different types of biometric markers for some time. TikTok's "For You" page, for instance, employs a sophisticated behavioral analysis system that tracks everything from how long a user watches a video before scrolling to how quickly they like or share content. Snapchat uses biometrics to deliver its digital image filters, which are, of course, very popular with children.

What's most unsettling is how these subtle markers can contribute to psychological profiles that ultimately feel deeply intrusive, especially when we've been unaware that they exist. The idea that a system could infer our child's emotional state or vulnerabilities without their knowledge or ours is incredibly problematic.

The truth is, this kind of data collection has been happening to our kids for years, and long before ChatGPT arrived on the scene. We can't just blame recent AI developments for these practices that have been deeply embedded in the platforms we use every day. But now that our eyes are starting to open to the possibilities, we can be more intentional about the choices we are making in what platforms to use.

CROSS-PLATFORM DATA AGGREGATION

Cross-platform data aggregation is a longstanding and often overlooked reality that should be a focus right now as well. This is where data from different services and platforms gets pooled together through advertising networks, data brokers, or corporate partnerships to build a profile that no single platform could create on its own. It's something that concerns me personally, as the implications reach far beyond any one platform or AI system.

This data collection ecosystem is largely unregulated by the US government, leaving Americans at risk of both commercial exploitation and cybercrime.[8] Location data can be particularly problematic, as it has been used to track individuals without their knowledge. As mentioned earlier, Texas is suing Allstate Insurance for having used location data to inform insurance rates. And while there have been calls to better regulate the industry that sells or "brokers" this information, there hasn't been much progress to date.

Even if a family carefully configures privacy settings on one platform, data from other services can be combined to reveal intimate details about a child's academic history, emotional well-being, household context, or personal vulnerabilities. These aggregated profiles are largely invisible to the families they describe and nearly impossible for adults to see or control. And this is precisely what happened to my friend, when her daughter was forced to watch commercials for anxiety medication on a streaming service after privately seeking help.

While AI technologies can make data aggregation more robust and efficient, it's not new. Yet in discussions with my family and friends on this topic, most don't realize the scale of data collection or exclaim that they "have nothing to hide." But it's important to remember that often there is a bigger picture that can be stitched together using many seemingly "meaningless" bits of information. Think of it like a puzzle. A single blue puzzle piece tells you nothing. It could represent a blue sky, water, or someone's shirt. But when combined with hundreds of other pieces, that blue fragment becomes part of something more detailed. And it might reveal a lot more than we ever expected or intended.

Cumulative Effect of Data Collection Methods

All of these different data collection approaches work together, accelerated by new technology, to create an extraordinarily detailed portrait of you and your child's lives. As I shared with the puzzle analogy, we must see the bigger picture and respect the power of every data point, no matter how seemingly "small."

When AI systems combine all of these data fragments, they don't just understand what your child likes or needs—they might predict what they'll want tomorrow, how they'll react to challenges, and which influences they may be most swayed by. This predictive power is what makes AI systems increasingly useful, from accelerating learning to identifying academic problems before they become serious. But these same predictive systems also create vulnerabilities that I'll explore throughout this book.

The challenge isn't just what gets collected, but how these detailed profiles can persist indefinitely. This concern isn't new either. In the 1970s, national security experts introduced the concept of the "mosaic theory," based on the idea that even small, seemingly unimportant pieces of information might be combined to create a complete picture of a person.[9] It could put at risk those in government with sensitive roles. But the principle applies to all of us. And unfortunately, once a detailed portrait of us exists, it's nearly impossible to deconstruct or erase.

This is why understanding the "permanence" of digital data is so crucial to the discussion of data privacy and our kids. Once information is sucked into these AI systems, it's nearly impossible to get it back, so we need to consider what leaves our control in the first place and what our expectations are for its use.

How Data Permanence Problem Has Worsened

Unlike previous technologies where you could correct digital mistakes by deleting files or changing settings, AI training data works differently. When your family's data contributes to training an AI model, that information may influence the system's behavior long-term, even if you later decide to delete an account or change your privacy settings.

This doesn't mean your child's name and photo are floating around AI systems forever. Most reputable AI companies work to anonymize training data, removing direct identifiers and personal details. But the patterns and insights derived from that data can still become part of how these systems operate. And as we just discussed, even anonymized data can be worrisome when combined with other information.

What makes this particularly frustrating is that we often agree to website and app terms and conditions without fully understanding what we're signing up for. Those privacy policies we click through frequently include vague language about data being used "to improve our product or service." What doesn't register in our minds is that "improve" frequently means our information becomes a permanent part of the system's foundation.

This reality makes our privacy decisions more significant than they used to be. Choices we make today may have lasting implications for our children as they grow into adulthood. And once data is baked into an AI system, it's really not possible to unravel. Trying to have this information deleted or removed is like trying to extract an egg from an already-baked cake. All the ingredients—the flour, eggs, oil, and sugar—have been transformed into something entirely new. You can't simply pull out the original components, and it's the same with AI datasets.

When you hear people talking about regulating "training data," this is what they mean. Schools, in particular, need to ensure that data collected for one reason doesn't end up being used for an entirely different purpose. Helping schools police this type of new risk—which, even with regulations, is incredibly difficult to do—is another reason for families to get involved with AI policy creation and enforcement.

A More Lasting Digital Footprint

AI systems don't simply store individual pieces of information that can be retrieved and deleted later. Instead, they analyze patterns across millions of data points and encode these insights into complex mathematical models. Your child's interactions and behavioral patterns can become part of a system's "knowledge" in ways that are distributed throughout the entire model.

You might think, "Well, what's the harm? It's anonymous information." But as we've just reviewed, even "de-identified" data can risk being connected back to an individual through the "mosaic theory" effect. Remember the point I made earlier about Meta using public Facebook and Instagram photos for AI training purposes? Even if you delete your account later, those images may have already contributed to the systems that power AI tools, and this contribution can't be reversed.

For families, this creates a new type of digital footprint that extends far beyond traditional concerns about photos or posts that might resurface later. Even anonymized patterns about learning styles, interests, or developmental stages could potentially influence the educational content your child encounters or the opportunities they're presented with in the future.

This is where we need to think about tradeoffs. AI systems that understand how children learn can provide better educational support. But that same capability should change how we think about every piece of information our family shares online. And often it's not the educational apps that pose the biggest risk—it's our social media content, the websites we use and then forget about, or documents we upload to shared drives without a second thought. The list goes on. We've been too loose with our information without considering the long-term consequences. But now is the time to change our thinking about all of this.

Snowball Effect of Data Collection Now

What makes data permanence even more complex is how our information moves through an interconnected web of partnerships, acquisitions, licensing agreements, and public datasets. When we agree to a single company's terms and conditions, we typically don't realize that the data can also be shared between partners. It's just one of the hidden complexities we opt into without realizing the scope of this type of permanence.

Here's what that looks like in practice: A child's anonymized interaction patterns from an educational app might contribute to speech recognition models used by multiple companies. That same data could be incorporated into recommendation systems, licensed to third-party developers, or even released as part of publicly available training datasets that anyone can download and use. Once data enters these public repositories—often collected under the banner of "research" or "open-source AI development"—it becomes nearly impossible to track or control who accesses it or for what purpose.

We're not talking about AI system training happening in isolation within a single company's walls. Your family's data can live across multiple organizations simultaneously. Research institutions, tech companies, academic projects, and startups all share and build upon the same datasets. A voice recording initially collected by one service might end up training models for dozens of different applications you've never heard of and would never have chosen to support.

Companies regularly merge, are acquired, or change business models in ways that can significantly alter how our data is used. A small educational startup with careful privacy practices might be acquired by a larger company with entirely different approaches to data use and sharing. When that happens, the data you trusted to one organization suddenly operates under new rules, new partnerships, and new purposes.

The issue of data permanence isn't just about one company keeping your information—it's about how our data can live on through many iterations of partnerships, public releases, and sharing arrangements we never explicitly agreed to. This ecosystem effect multiplies the risks and makes true data deletion virtually impossible once information enters the broader AI training pipeline.

Teaching Kids About "Forever" Data

The data permanence challenge reinforces why family conversations about digital privacy matter so much. Rather than making individual rules about each new AI tool, consider creating overall family guidelines about what types of information you're comfortable sharing and what you all agree to avoid.

The scope and scale of the data required to train AI systems should also help us understand what we're up against in trying to delete or retrieve previously shared information.

Generative AI systems like ChatGPT are trained on trillions of data points today. Just to launch ChatGPT in 2022 required the analysis of roughly 300 billion words, and we are far past that now.[10] Imagine trying to find and delete a single data point belonging to your family somewhere in that massive dataset. It's simply not possible.

This is why regulations like the European Union's "Right to Be Forgotten" law are nearly impossible to enforce in this new AI age.[11] Current regulations struggle to keep pace with the reality of digital data collection, which means we need to address these challenges first and foremost at the point of sharing the information—not when we try to erase it later.

Helping your children understand that their interactions with AI systems may influence how these tools work over time is key to motivating them to engage thoughtfully. This isn't about creating fear but about building the critical thinking skills they'll need to make their own privacy choices well into the future.

Most importantly, remind your kids that this is an evolving landscape. The AI industry is actively working on better privacy practices, and regulations are starting to catch up to the technology. By staying informed and making thoughtful choices, families play a part in shaping what happens next. And one of the most powerful ways we exercise that choice is through the consent we provide when signing up for digital products and services. Understanding what it really means to "opt in" to the platforms we use each day is precisely where we're headed next.

Shannon Kimberly Edwards

Understanding Digital Consent

When most of us think about giving permission for data collection, we might not fully appreciate that we do this by checking a box after scrolling through—and not reading—a lengthy and convoluted document. It's fair to think of this simply as an annoying formality because there seems to be very little we can do anyway.

But these policies are important, and consent in the AI era is now far more complex than ever before. The terms are often very loose and sweeping, flexibly "interpreted" to give consent for AI training. So we do need to understand what we're getting ourselves into from the start.

The scale of this challenge was staggering even before new generative AI tools entered the picture—and we've long given up bothering. Research shows that more than 80% of Americans feel they have no control over the data that companies collect about them.[12] Even more telling, only 22% of people say they've ever read a privacy policy all the way through before agreeing to its terms.

And who can blame any of us? One study found that reading the privacy policies of the 96 websites a person typically visits in a month would take 46.6 hours.[13] That's literally longer than a typical work week. Of course we don't do this and never will, but it's exactly this effort to outmaneuver users with ridiculously long policies that we should be focusing on and calling out.

Even if you do understand what you're opting into, when you click "accept all cookies" or agree to an app's terms and conditions, you're typically consenting to sharing data with other companies as well—often an unknown list of partners. These might include advertising networks, data brokers, analytics firms, and AI training companies. Agreeing to use a

124

simple educational app can actually mean consenting to a complex web of data sharing that could include even unrelated companies and services.

These terms and conditions are intentionally left loose to accommodate a long list of partnership opportunities. While it would be difficult for companies to update these policies more regularly, it's consumers who are left dealing with the implications of this loose language and broad permissions.

How We Consent to "Continuous" Use Our Data

Many AI systems also interpret your continued use as ongoing permission to collect and analyze your data in increasingly sophisticated ways. When your family uses voice assistants, smart home devices, or educational apps, the systems often assume you're okay with them learning from your interactions and improving their services, even as their capabilities evolve.

This "implied consent" model is particularly tricky because AI capabilities develop rapidly. An educational app that initially analyzed only academic performance might later add emotion recognition, behavioral prediction, or social relationship mapping—all under the umbrella of your original agreement.

Interconnectedness of Consent Within Families

Consent becomes even more complicated in family settings where multiple users, ages, and privacy preferences intersect. When families create accounts that children use or when family members share devices, it can be unclear who has actually consented to what data collection and when.

Children themselves can't legally provide consent for apps and platforms, yet they're typically the primary users generating the most valuable data. This creates a gray area where parents might consent to basic educational features without realizing they're also authorizing comprehensive behavioral analysis of their children.

When multiple children use the same account, their data profiles can get mixed, creating algorithmic assumptions about each child based on their siblings' behavior. The same can be true of adults who may find that their privacy is compromised based on information gathered from their children.

What Data Privacy Protections Exist

I know you're thinking, "Surely there are laws to protect us from all this?" And yes, there are some, but as we've covered previously, the regulatory landscape is a patchwork of protections that depends heavily on where you live.

COPPA: THE FOUNDATION

As I mentioned in Part One, COPPA is the main federal law protecting kids online and is very data privacy focused, requiring companies to get parental permission before collecting data from children under 13. This sounds good in theory, but as we've covered, it struggles to keep up with how AI actually works in practice.

The consent process often fails to adequately inform families about how AI systems will actually use their children's data. When a privacy policy mentions "machine learning algorithms" or "automated decision-making," most of us lack the technical background to understand what this means for our child's privacy.

AI capabilities also evolve so rapidly that companies can expand their use of children's data in ways that weren't anticipated when parents first gave permission. COPPA's framework continues to struggle to keep up with these changes in data use.

1974 FEDERAL **FERPA** ACT

The Family Educational Rights and Privacy Act (FERPA) is supposed to protect student records and, by extension, all of their data privacy in schools, but this decades-old law can't keep up.[14] First, it was passed before the advent of the Internet. It contains assumptions about student data that just don't fit today's reality.

When schools use AI-powered educational tools, your child's academic data often flows to commercial companies for analysis and system improvement. FERPA's consent framework doesn't adequately address how this educational information can become part of broader AI training datasets. Parents might assume their child's school records are private, but those same learning patterns could be contributing to commercial AI development in ways FERPA never anticipated.

The law also struggles with the permanence problem we discussed earlier—once educational data becomes part of AI training, it's nearly impossible to remove, even if parents later object to how it's being used. And unfortunately, a 1970s-era law just can't accommodate this reality.

MIXED BAG OF PRIVACY-RELATED STATE LAWS

Several states have enacted more comprehensive privacy legislation, but these laws frequently struggle to address how personal data becomes AI training material. For instance, California's Consumer Privacy Act gives residents the right to know what personal information a company collects, request its deletion, and opt out of its sale.[15] But even these strong protections don't adequately cover data used for AI training purposes.

It can also be frustrating that your zip code determines your privacy protections as well. Some companies choose to apply the most rigorous standard (usually California's) to everyone, but this approach isn't universal (and frankly, a state-by-state approach to a boundaryless internet is ridiculous).

GDPR Model in Europe

While US families aren't directly subject to European Union law, the General Data Protection Regulation (GDPR) provides the world's most comprehensive framework for data protection.[16] GDPR requires explicit consent for data processing, gives people the right to an explanation for automated decision-making, and includes a "right to be forgotten" provision (as I've mentioned already). For children under 16, GDPR requires parental consent for data processing, extending protections beyond COPPA's 13-year age limit.

Many global tech companies adopt GDPR standards across their platforms rather than having to maintain separate systems for different regions. When evaluating AI systems for your children, it's still worth exploring whether a company extends these protections to all users regardless of location. Understanding GDPR can also give you context around what privacy laws should be strengthened in the US.

Opportunity to Build on Current Protections

Current legal protections provide a foundation, but they fundamentally struggle with how AI systems work and use personal data today—which is why understanding digital consent at the point of use is critical. The fact is, privacy consideration starts with us, and there are plenty of companies to choose from, with many committed to protecting our children's information.

Overall, it might feel like a lot to wrap our heads around, but hopefully you're starting to see how much power we have and how much we ultimately give away when we don't pay attention. Your choices about which platforms to use, what data to share, and how to engage with AI systems collectively influence how these technologies develop and grow. Companies pay attention to user preferences and concerns, especially when they're expressed consistently across many families.

The goal isn't to become a privacy law expert or spend those 46.6 hours reading every policy attached to every platform you use. It's to understand enough to make informed decisions that align with your family's values. Understanding your part in protecting your family's privacy is also critical when we consider the even larger risks, including data theft from cybercriminals.

How AI Can Power and Amplify Cybercrime

Imagine your school's principal receives an email from the district superintendent that looks identical in every way to past emails—same tone, style, and official signature. The email asks that student data be sent immediately as an attachment in response to the email. Does the principal comply? What if they do a video call to verify and unknowingly speak with a convincing AI-generated deepfake?

This scenario isn't science fiction. It's happening right now and represents a thoroughly different category of threat than the commercial data collection we've been discussing in this chapter or any pre-AI era cybercrime-type events.[17] These sophisticated impersonations can make it incredibly difficult to distinguish between what's real and what's artificially generated, and even the savviest of us are getting tricked as a result.

In the past, fraudulent communications were relatively easy to spot with their signature bad grammar, wonky graphics, or suspicious links that led to stolen financial information or hacked accounts. But AI has changed the game entirely. Not only are fake communications more convincing, but they can also result in a much broader spectrum of harm, from damaging one's reputation to manipulating relationships and causing real psychological damage.

The scope of AI-powered fraud is already staggering. Just this year, the US Department of Education discovered that $90 million in financial aid had been secured by students who don't even exist.[18] These "ghost students" were created by AI technology and then enrolled in online courses with the aim of securing federal funding—and it worked.

For children and teenagers, these threats can be particularly challenging because they occur during crucial developmental periods when peer relationships and social standing profoundly impact mental health and future opportunities. But it's important to remember that understanding these risks doesn't mean living in fear. It's similar to any safety-related issue we navigate with our kids, and our approach should be similar.

Why Kids' Data (and Identities) are Most at Risk

Children's data is of particular interest to cybercriminals for identity theft because it is, as security experts say, "clean data."[19] This means that your child's information comes without the complications that adult data carries.

Think about it from a criminal's perspective. Your child likely doesn't have a credit report, employment history, or existing financial accounts that would trigger fraud alerts. Their Social Security number hasn't been used to open credit cards or take out loans. This clean slate makes children's identities perfect for sophisticated fraud schemes that can go undetected for years.

Bad actors also don't necessarily need to hack into school systems or trick kids into giving up information directly. They can buy this data from the same places legitimate companies sometimes source it, such as via data brokers. These groups collect and sell information from various sources, including educational apps, games, and social media platforms that may not have adequately protected kids' data in the first place.

Unfortunately, the complexities don't end there either, because children's data doesn't exist in isolation but can reveal information about parents, siblings, household dynamics, and even family values and beliefs—extending the fraud far beyond an individual child.

How AI Has Changed Cybercrime Landscape

Creating convincing fake content intended to deceive no longer requires technical expertise or expensive equipment. AI systems can generate realistic images, videos, and text using freely available applications and minimal source material. So instead of the much-joked-about "Nigerian Prince" emails, we are instead looking at sophisticated outreach with content that seems real.[20] And thanks to chatbots, criminals can both collect the type of information about a target that would make one think it's legitimate and sound far more sophisticated in any phishing communication.[21]

The repercussions of fraud can be a lot more widespread too. Traditional identity theft focused primarily on financial exploitation, such as stealing credit card numbers or bank account information. AI-powered threats can more convincingly affect a person's reputation, relationships, or psychological well-being. The reason these attacks can feel more raw is that they often use familiar information or connections to people and situations.

When it comes to full identity theft and impersonation, this type of AI-generated fake content can become part of a child's permanent digital footprint as well. It can influence their future educational and employment opportunities, harm their credit, and create lasting issues that continue well into adulthood. Even when fake content is eventually identified and removed, it may have already caused real or lasting damage.

So what makes AI such a powerful tool for fraud? Unfortunately, it's because AI can be inserted every step of the way, from making phishing outreach seem personal or professional to the way AI can amplify the content or automate delivery. AI both supercharges the crime and increases the value of its bounty. Here's what to look out for:

AI-Era Tactics of Cybercriminals

The world of identity theft, impersonations, and other fraud looks very different now, with AI creating and amplifying threats that didn't exist even a few years ago. But when you understand the different approaches, you can spot red flags faster.

FAKE AND DECEPTIVE CONTENT

Of course we now all know that AI can generate realistic photos, videos, and audio recordings of people saying or doing things that they never said or did. But the same tools that are often used for fun or utility, like creating professional headshots, can also be misused by people with bad intentions. For children, this means that with just a few photos from social media, bad actors can create very convincing fake content. This fake imagery can appear to show kids in compromising situations or expressing views they don't hold.

Unfortunately, images shared through school platforms or with close friends aren't necessarily safe from this type of manipulation. High-quality images, like school yearbook photos, are valuable for their ease in creating convincing fakes.

In fact, schools in Korea are reportedly abandoning yearbooks for fear of offering any more fodder for deepfake creation.[22]

We don't need to abandon those yearbooks just yet, but it's worth asking the company that takes the photos, for example, to delete any stored images of your child (some have policies to never delete unless asked). It also helps to know what your school's policy is on image sharing, including on social media platforms. Overall, it's just further evidence of how much more careful families need to be now.

TARGETING OF INDIVIDUAL VULNERABILITIES

We discussed earlier the way AI can facilitate data aggregation and create a "picture" of who we are. Unfortunately, this is something that criminals now make ample use of. They can take information from multiple platforms and create detailed profiles that no single platform alone could generate. A child's gaming habits, social media posts, academic performance, and family photos can be integrated to create a detailed psychological profile to be ultimately abused.

This is the type of information that can make phishing attempts so believable. It also creates a double threat where AI both collects detailed personal information and then uses that information to create more sophisticated fake material. It can be so convincing as to make kids vulnerable to predators, scammers, or other malicious actors.

AI systems can also be used to identify children who are struggling, lonely, or looking for validation—making them easier targets for people with bad intentions. This isn't an entirely new tactic, of course. Social media and gaming platforms have always attracted people who prey on vulnerable kids. But AI makes it much easier to systematically identify and target children who might be most susceptible to manipulation and then weaponize what a criminal might know about them.

INITIATION OF FAKE RELATIONSHIPS

One of the most disturbing ways AI can be misused against children involves deception around relationships. Since our kids naturally build friendships and connections through online channels, the idea that someone could exploit these relationships is particularly troubling. And again, criminals "posing" as kids or someone trusted online is not new. What is new is the sophistication with which they execute these scams.

AI-generated fake profiles can impersonate children's friends, family members, or romantic interests to extract personal information, manipulate emotions, or encourage risky behavior. These fake personas can be sophisticated enough to maintain long-term relationships that feel completely genuine to the target.

The more familiar the profile too, the more kids share, and it becomes a vicious cycle. They can be convinced that they are talking to a family member or a peer at school via chats or other modes of communication. What it requires from us now is a new take on not talking to strangers.

Impact of Deception on Families

These AI-enabled identity threats are particularly damaging because they exploit fundamental human needs for authenticity, trust, and social connection at a time when our kids are most vulnerable. Identity theft affects more than just finances or convenience. Victims can experience real physical symptoms as a result, making this a health issue as much as a security one.[23]

For children specifically, these attacks can cause confusion about what's real, loss of confidence in online environments, and fear of using digital tools that are essential for their education and social development.

But the most important thing to remember is that we've got

this. We know the risks; we just need to get smarter about the modes of delivery. Awareness is the first step to better protecting our kids from their effects, and there are concrete actions families can take both to prevent these situations and to respond if the worst does happen.

Unfortunately, even with every precaution taken, identity theft still happens. And AI doesn't change the need to keep an eye out and respond quickly. To help understand the steps, I've added resources and a check list on my website, aiforfamilies.com. There are also a number of consumer protection organizations that can support families in taking immediate action. If we act fast, the damage can be mitigated, but hopefully, with your newfound awareness of data issues, avoiding many of these problems out of the gate will be easier.

Key Chapter Takeaways

After working through all the information in this chapter, you might be feeling equal parts concern and—hopefully—determination to make changes in how your family considers the collection of their data. The concern is understandable. We've covered some serious territory, from sophisticated data collection to identity theft and manipulation tactics. But knowledge is power, and understanding these issues is the first step toward confidently protecting your family.

AI systems have indeed changed the privacy and identity threat landscape in ways that are challenging to keep up with. Unlike traditional risks that were easier to spot, modern threats often hide within familiar platforms and interactions, making them harder for both families and children to recognize.

And it's not just the chatbots we're coming to terms with,

but the way AI technologies can systematically amplify issues in the backend as well. But throughout this chapter, I hope you've noted that even the smallest decisions we make about our privacy can make a substantial impact down the line, especially when done together. Simply adjusting one setting or rejecting a platform or a company's terms and conditions is power. The fact is, companies need our data and want us to like their products and services. The key is recognizing the value in our information and treating it as precious.

AI's Overwhelming Need for Data

In this chapter, we've explored how our data is collected through apps, games, emails, social media platforms, and more —in ways that can create permanent digital records that are difficult to eliminate. The resulting problem of data permanence shouldn't make us feel defeated but instead more considerate of the data sharing choices we make.

The scope of collection has expanded far beyond what most of us can wrap our heads around, especially the way our kids' data is sought after by criminals. A child's gaming habits, academic performance, social interactions, and family photos can be integrated to create comprehensive profiles that reveal far more than any single piece of information would suggest. And creating convincing fake content no longer requires technical expertise or financial resources either.

On the darker side of data collection is an ecosystem— fueled by data's new elevated value to AI training and the AI systems themselves—that can make stealing data much easier. Unfortunately, kids are at the epicenter of it all. According to the Center for Internet Security's (CIS) 2025 K-12 Cybersecurity Report, 82% of the 5,000 schools polled experienced a cyber threat over an 18-month period.[24] These are staggering numbers that will only grow if we don't take the threat seriously and help our schools to better prepare.

The Data Ecosystem's Impact on Families

The challenge we all face is distinguishing between applications that genuinely support child development and those that primarily serve commercial interests or put our kids at risk of fraud. This means understanding not just what data is collected but how it's used, who benefits, and what long-term implications exist.

This makes providing "consent" a much more serious activity than we've given it credit for. The terms we agree to often grant broad permissions for data use that extend far beyond the immediate service. Children's information gets shared across platforms and used to train systems in ways that weren't clearly explained when we first signed up.

But protection begins with awareness, consideration, and intention. It means being able to identify the terms that are red flags or that protect us in company policies without spending hours reviewing the documents. It also means helping children recognize potential issues—overly personal questions from AI chatbots, requests for sensitive information, or interactions that make them feel uncomfortable.

If identity theft does occur, remember that while it's serious, it's also manageable. Document everything, secure accounts immediately, work with financial institutions and credit agencies, and don't hesitate to seek support for your child if they're struggling with the impact.

According to my friend Kirsten Co, a master's degree candidate in Global Security, Conflict, and Cybercrime at NYU, the most important thing families can do in the case of a data breach or fraud is simply to take action. While child identity theft is serious, it's manageable, she says, with the right steps. "Many families successfully recover from these incidents, and acting quickly can significantly limit the damage," said Kirsten. "The key is to take as many precautions in the beginning as possible, and if the worst happens, don't shy away from taking the steps to address the crime."

Most importantly, maintain open dialogue with your family about online experiences. Children should feel comfortable sharing when something makes them uncomfortable, confused, or upset without fear of losing device privileges. This requires responding with curiosity rather than judgment and helping children process their experiences rather than simply restricting access.

Advocating for Better Data Protections

The current lack of comprehensive data governance has been compared by UNICEF to the unregulated nature of financial services that led to the 2008 recession.[25] Just as stronger financial regulations ultimately benefit the entire economy, better data protection standards can benefit everyone—families, children, and businesses alike. Advocating for better data privacy protections isn't about stopping or slowing innovation. It's about ensuring that innovation serves everyone's interests.

Think of privacy protection as an ongoing family conversation rather than a one-time discussion. As technology evolves and your children develop, your approach will need to adapt. By teaching kids to question, verify, and think critically about online interactions, you're giving them tools that will serve them throughout their lives. The key is building awareness, maintaining open communication, and developing critical thinking skills that will serve your family regardless of how these technologies change.

As you start to think about how you can better advocate for data privacy at the federal, state, and local levels, there's another piece of this puzzle to consider: the economic value that companies extract from our personal information. Even with well-intended legal or safe data collection policies, we need to start asking if this exchange is even fair at all. Do we benefit in the same way that the companies learning from us do?

So to start considering these questions, we first need to better understand how companies profit from our information. Hopefully, by shifting our thinking regarding this dynamic, we can start to consider the true power of the "currency" we all hold in our data.

Chapter 5
Why We Should Think of Our Data as Currency

For far too long, we've avoided understanding in detail the value we receive from the digital platforms we use and whether the exchange we make for our data is a fair one. There is no easy answer to this question, of course, nor will it be the same for everyone. But there is a very robust, and often imbalanced, exchange going on, and hopefully you are ready to think about what it means for your family, especially now that AI demands more of our information to grow.

This is something my former colleague Nico realized after a few months of his son using a free homework helper app. Up until that point, he thought he was being smart about his family's digital privacy. Nico taught his 14-year-old son to never share personal information online, avoid posting too many photos on social media, and carefully review app permissions before downloading anything new. But after Nico and I looked together at the app's privacy policy and talked about what information it was collecting, Nico realized he'd been missing a much bigger picture.

"When I started to look more closely into what we were

using, I realized I needed to better understand whether we were giving away more than we were getting back," Nico told me. "I was so focused on teaching my child to avoid strangers online and not share images on social media that I completely missed how much information I was letting him hand over to companies that could profit from these details."

Nico's not alone in feeling this way. Academics at the University of Pennsylvania conducted a comprehensive review of American attitudes toward data sharing and consent and found two opposing extremes.[1] First, 91% of Americans surveyed agreed with the statement, "I want to have control over what marketers can learn about me online." But 79% also agreed with the statement, "I have come to believe that I have little control over what marketers can learn about me online." That's a lot of people who want control but feel like they don't have it.

As I've shared previously, I do not believe we are powerless. In fact, the biggest risk we face is in remaining indifferent. But the good news is that it doesn't take much work to shift the power dynamic. Even just the act of being more intentional with our data sharing can create the type of healthy "friction" that slows down the one-way flow of our data. It starts by taking a look at the bigger picture of how this data system works.

Nothing in Our Digital World is Ever "Free"

The starting point for any discussion about the economics of digital data is taking a closer look at those "free" apps and services we never really question. It's like if a neighbor offers to watch your kids for free, but then you find out they took notes about the toys your kids played with and sold them to a local marketing company. You thought your neighbor was being generous, but they just found a different way to get paid.

That's essentially what's happening with billion-dollar tech companies. They give you services like email and social media for "free," but then they document your every behavior, follow you around the web, and use all of this data for various, mostly marketing, purposes—including today to train AI systems.

And now too many companies are crossing the line in what information they are taking from unknowing consumers as well. According to Common Sense's 2023 Kids Privacy Report, of the more than 200 apps and platforms they looked at, 73% monetized kids' and families' personal information in ways not apparent to users or legally compliant.[2]

This data collection ecosystem also extends far beyond any single app or tool we may use. The educational app shares learning data with textbook publishers. The gaming platform shares engagement patterns with toy manufacturers. The social media company shares demographic information with college recruiters. The web of data exchange expands on and on.

You may be wondering if this is really any different from the regular online advertising world we've resigned ourselves to after years of popup ads and other marketing intrusions. The answer is yes—it's much more complex and far less transparent. We've moved beyond simple ad targeting into something bigger.

Harvard professor Shoshana Zuboff calls this newly expansive and invisible data-collection ecosystem "surveillance capitalism."[3] What she means is that we are now part of a sophisticated market that trades on behavioral data for the purposes of training increasingly sophisticated machines. By simply going about our everyday lives online, we become profitable assets. And we are mostly none the wiser about it.

What You Will Discover in This Chapter

We have far more power here than we might think. We are integral to the creation and maintenance of this AI ecosystem, and that gives us more of a say than even just a few years ago. Just like we teach our kids to manage their allowance, we should teach them to be selective about which companies get access to their personal information.

You've probably heard the saying, "If you're not paying for the product, you are the product." When we understand the data required to build and improve AI systems, we can start to rethink those services where we're treated less like customers and more like sources of valuable information.

We'll also look in this chapter at the pressures coming from the technology industry, as AI innovation strains an already-fragile dynamic between consumers, government, and Big Tech. And we'll talk about why diversity is so important to ensuring that the largest number of people benefit from the AI systems being created.

The good news is that understanding how our data becomes profit is the first step to taking back some control. Once we see the exchange clearly, we can start making different choices about where our information goes and what we get in return. For families, that means learning to treat our data like the valuable currency it actually is.

How Personal Information Becomes Profit

You may still be wondering how AI changes the value dynamic when our information has always been used to sell us products and services. Well, AI shifts things in three important ways.

First, AI makes it easier to collect far more detailed data than ever before. We covered this in the last chapter, but think of it like the difference between a company knowing you clicked on a shoe ad versus having a detailed record of why you need new running shoes, what your budget is, and what your running goals are. And this richer data is now coming from more places than ever—not just your web browsing, but your conversations with chatbots, your voice commands, and your everyday app interactions.

Second, AI creates entirely new products and services that companies can monetize. They're not just using our data to show us better ads anymore. Now companies are building AI-powered services and even potentially licensing the technology to other businesses. Our data has become an opportunity for commercial entities to grow and expand in new ways.

Third, data is core to AI development. Without our data, no company would be able to build new AI systems at all. We're not just the audience anymore—we're the fuel that makes the entire AI economy run.

When you start to think of your data as having this type of real monetary value attached to it, you can see the power we actually have in the exchange. You don't need to become a data scientist or understand every technical detail. You just need to recognize that what you're giving of your data has measurable value. So let's look at who is making money off us and how.

The Prediction Economy in Full Swing

Every app your family uses doesn't just collect data—it builds a profile to help "predict" future activity, wants, and needs. When a child uses an app that captures how long they hesitated before answering a question or which problems made them start over, the app creates a valuable picture of that user, especially when it's linked to an adult decision maker.

Think about it this way. If a company could predict with 85% accuracy which kids would struggle with algebra next semester, who would want that information? Tutoring companies, for one. Educational publishers might buy that data to offer supplemental workbooks to specific school districts. Test prep companies would want to know which families to approach with "algebra bootcamp" offers during summer break. And which families could afford to send their kids to more expensive programs.

Many families of high school students know this reality well as their children start to think about college. Recruitment firms pay premium rates to get their hands on family and student information—from test scores and subjects studied to clubs joined and sports played. Financial services companies use this data to market student loan products. The list of organizations interested in this information is long, and it includes some of the most trusted names in education.

While much of this data trading has gone on for years, AI now supercharges the process and creates new opportunities for collection and analysis. The scale and sophistication have changed dramatically. Take the College Board, which administers AP exams and the SATs. The organization has been accused of using deceptive tactics to profit from the data they share with colleges.[4] Advocates such as the Parent Coalition for Student Privacy estimate the national database of student information to be worth at least $100 million a year to the organization.

But it's not been without consequence: In 2024, New York State's Attorney General settled with the College Board on behalf of hundreds of thousands of students whose data had been "licensed" to third parties without proper consent.[5] The settlement was a victory, though the economics remain frustrating. The $750,000 fine sounds significant until you consider the estimated value of their full database. When the fines are small compared to the revenue these companies generate, it's hard to see the incentive to stop the practice.

Growing Influence of Data Brokers

We've talked a bit already about the business of "brokering" our data and how it's become a thorn in the side of many consumers. What's important to understand now is how incredibly profitable these businesses have become—and how fast they are growing. The industry is currently valued at around $270 billion and is expected to reach $473 billion by 2032.[6] The industry is also as opaque in its operations as it is profitable.

Data brokers specialize in collecting information from multiple sources and then selling comprehensive profile information to hundreds or even thousands of other businesses. And as I shared previously, the way the data gets combined, passed on, and then combined again makes it nearly impossible to identify the source or have it deleted.

Your child's educational app data gets combined with your social media activity, shopping patterns, location information, and health searches to create detailed family profiles that are sold and resold without your knowledge. The same family profile might be purchased by college recruiters, insurance companies, marketing firms, and financial services companies, with each buyer getting the specific data points they need to target your family more effectively.

Financial institutions increasingly use this data to make decisions about credit offers, insurance rates, and loan terms. The Federal Trade Commission (FTC) took action against

General Motors (GM) in 2025 or selling OnStar geolocation data and driving behavior information to third parties—including consumer reporting agencies—without consumers' consent.[7] And as I've already shared, Texas's Attorney General has sued Allstate for using location data to adjust insurance rates. These examples are just the tip of the iceberg, and the practice shows no signs of slowing down.

High Value Placed on Location Data

Location tracking data deserves special attention when it comes to families. Many apps turn location access on by default because this information is extraordinarily valuable for targeting advertisements and building behavioral profiles. It's also incredibly lucrative for data brokers.

In early 2025, a global consortium of news organizations analyzed a dataset from a data broker that contained 47 million "Mobile Advertising IDs," covering 380 million location data points in 137 countries.[8] The information came from a staggering 40,000 individual apps, including popular names in gaming, dating, shopping, news, and education—apps that many families and their children use every day. The implications are serious and range from government surveillance to the physical safety of children to the privacy of those seeking refuge from abusers.

This is why it's worth checking the location settings of every app you use and turning off what you don't need. Many app settings default to opt-in, so you need to manually adjust them to protect your family's information.

Data brokers have been operating in this murky space for a long time. In fact, in 2014, the FTC released a report detailing the growing data broker industry, identifying risks, and calling for better regulation—which has not happened in any meaningful way since the report.[9] More than a decade later, meaningful regulation still hasn't materialized. The difference now is that AI has amplified the value of this data exponentially. An innovation that relies so heavily on data needs vast amounts of it from somewhere, and data brokers are more than happy to supply it—putting all of us at greater risk.

The Race to "Train" AI

When AI companies talk about "training data," they mean your stories, photos, voices, schoolwork, shopping patterns, and the everyday digital footprints that families generate—all used to "teach" AI systems how to work. This raw material powers AI systems, which in turn power all the industries using AI.

The scale of this market is staggering. In 2024, the global market specifically for AI training datasets was worth between $2.6 billion and $2.9 billion. By 2030–2034, estimates suggest it could balloon to between $9 billion and $18 billion. [10]

What's also important to know about this new "training data industry" is how much it also depends on scarcity. Common, everyday data is easier to come by. But high-value data—like a child's speech development patterns, unique medical images, or multilingual text in less common languages—commands premium prices because it's harder to find. These rare datasets help AI companies reach new markets or improve accuracy in ways that generic data can't.

These numbers represent the real monetary value of your information to a growing industry. When you understand this, those constant requests for your data start to make more sense. Whether it's a good-faith request from a company trying to improve its AI system or a more sophisticated phishing attempt from a bad actor, they're all after the same thing—your valuable information. The insatiable need to train AI systems is driving this surge in requests, and often we don't even realize what's happening.

The Data Owner Mindset

Understanding these economic realities doesn't mean you should avoid digital tools. It's about reclaiming our power. Think of it like understanding how grocery stores make money —it doesn't stop you from buying food; it just makes you a smarter shopper.

It's also not to suggest we move to a pay-for-data model, as some have proposed, since that risks commodifying our privacy and weakening any protections that exist.[11] Instead, it's about cultivating an "ownership mentality." When you know how the data economy works, you can approach digital services with a clear sense of what's best for your family.

You have more influence here than you might realize, and understanding how the system operates is essential to using that influence effectively. Once you see how our information becomes profit, you can start acting like a customer instead of passively serving as the source of AI development—or worse, being treated simply as the "product" itself.

How to Be the Customer, Not the "Product"

Choosing a technology product or service should be as serious a decision as any other high-impact purchase we make. Think of how you might go about buying a car or a TV. You might not be able to describe in detail how the engine of a car works or explain the TV's display technology, but you can still make a confident, informed purchasing decision in both cases. The same is true for digital products—once you understand what to look for, evaluating these services becomes much more manageable. And you don't need to fully understand how the technology works to make good choices.

When you approach digital services as a customer rather than just a user, everything changes. A confident customer knows their rights, has higher expectations, and is more discriminating about whether a product or service is worth their time and data. The shift is simpler than it sounds. You just need to recognize when a company starts treating you less like a customer and more like the "product" they're selling to someone else.

The Power in Being a "Customer"

When you're genuinely the customer, companies prioritize your satisfaction, convenience, and long-term relationship. Think about your experience with a subscription service you pay for directly, like Netflix or Spotify. These companies invest heavily in user experience because they depend on your continued subscription payments.

As a paying customer, you have real power in the relationship. If you're dissatisfied with the service, you can cancel and choose a competitor. The company knows this, so they work to keep you happy.

Now, this doesn't mean subscription companies aren't using your data to improve their products or even selling insights to third parties. But because they rely on your subscription payments—and because there's enough competition that you can switch if you're unhappy—the power dynamic is fundamentally different.

This traditional customer relationship includes transparency about costs, clear terms of service, and respect for your time, attention, and right to privacy. Companies that depend on customer payments understand that treating you poorly means lost revenue and damaged reputation. They also need policies strong enough to operate across different states and countries, which often means better protections for everyone.

These companies make it easy to cancel, provide responsive customer service, and regularly improve their offerings based on user feedback. They want you to feel in control because your satisfaction directly translates into their success. When you start recognizing these patterns, you can better identify when a company is treating you like a customer versus when they're treating you like a commodity.

Compromises in Being the "Product"

When you're the "product," the company's primary relationship is with third parties rather than with you. This might include advertisers, data brokers, and other businesses that pay for access to you and your family's information. Your satisfaction becomes secondary to maximizing the value you provide to these paying customers.

We allow ourselves to be the product more often than we'd like to admit. When we access "free" platforms or services, we're usually agreeing to this arrangement. The telltale signs are clear once you know what to look for. Your needs, enjoyment, and satisfaction aren't the primary goals. Instead, companies optimize for engagement and data collection. They use sophisticated techniques to keep you and your children using their platforms longer, sharing more information, and developing stronger dependencies on their services.

Many families wrestle with this tradeoff when considering their child's social media use. Does connecting with friends via a social media platform provide enough benefit to justify the data, behavioral insights, and family details that these companies collect in return? It's a calculation that becomes clearer once you understand the true cost of "free."

Spotting the Difference in How We Are Treated

Learning to recognize when your family is the product starts with paying attention to how companies actually treat you. You might be surprised by what you discover when you start looking for these signs—or how often you accept poor service without questioning whether you should continue using a platform.

One major red flag is privacy settings that are impossible to find or adjust. If these controls are buried in complex menus, frequently reset to default settings, or explained in confusing language, the company is prioritizing data collection over your preferences. Similarly, if you can't easily delete your account— if it takes more than two clicks to find the option, or if the process involves multiple "are you sure?" screens designed to change your mind—you're probably dealing with a company that sees you as inventory rather than a customer.

Many platforms use what are called "dark patterns" to discourage making privacy-protective choices.[12] These are deliberate design decisions that make it harder for you to protect your data so a company doesn't lose this precious contribution of rich data. It's like a store putting expensive items at eye level while hiding the budget options on the bottom shelf. Watch for gray-shaded buttons versus bright-colored ones when trying to turn features off, or intentionally convoluted language that requires multiple readings to understand what you're actually agreeing to. If links to privacy policies are broken or inaccessible, that's another sign to reconsider using the platform entirely.

When Platforms Manipulate Rather Than Serve

Some platforms try to make you feel guilty for protecting your privacy. They suggest that enabling protections might significantly degrade your experience or disappoint other users. While tradeoffs between convenience and privacy are often legitimate, be skeptical of platforms that seem to punish privacy-conscious choices.

Pay attention to whether a platform has an imbalance of addictive features—infinite scrolls, intermittent reward systems, social pressure tactics, and notifications designed to pull you back repeatedly throughout the day. These are designed to maximize your engagement time, not your satisfaction.

Customer service quality tells you a lot too. Companies that profit from selling your data typically provide minimal customer service because resolving your problems doesn't directly impact their bottom line. If you can't reach a real person when something goes wrong, that tells you where you stand in their priorities.

Finally, notice whether the rules keep changing in the company's favor. Frequent changes to terms of service, privacy policies, and platform features that increase data collection or reduce your control are red flags. Companies that see you as the customer make changes primarily to improve your experience and typically protect existing users when policies shift.

Navigating the Relationship With Big Tech

With new AI technologies changing what's valuable and how our data is used, most platforms now use a hybrid revenue model. You might pay for a premium subscription while the company also profits from your data and attention. We see this everywhere—from streaming services to social media platforms with paid tiers.

This hybrid approach doesn't change the fundamentals of the customer-product dynamic—it just makes the relationship more complex. But here's what hasn't changed: there's still plenty of competition for your attention, data, and business. You can choose platforms based on what matters to you and your family.

Think about how we act in other industries. When we're unhappy with a restaurant or a grocery store, we stop going. We leave reviews. We tell our friends. Companies in these industries know that bad customer experiences damage their reputation, so they work to keep us satisfied. In the technology industry, though, we tend to shrug and accept poor treatment. That needs to change.

Understanding whether you're the customer or the product helps you make smarter decisions about digital engagement. This doesn't mean avoiding all services where you're the product—sometimes the tradeoff makes sense. But it does mean approaching these platforms with appropriate skepticism about recommendations, being more protective of your personal information, and maintaining stronger boundaries around usage time.

When you're genuinely the customer, you can engage more openly while still maintaining healthy digital habits. You have recourse when things go wrong, and the company has real incentives to prioritize your family's experience. As Big Tech companies pour more resources into AI development, the pressure on our data will only intensify. That's why recognizing these patterns now matters so much.

Better Understanding Industry Interests

Now that we've explored the ways our information drives company profit, let's look at the companies behind the push to monetize our data.

The tension between industry interests and consumers isn't new. Companies have obligations to their shareholders and need to deliver strong financial results. We want them to innovate and continue to offer competitive services and exciting new products.

The advocacy work technology companies do isn't necessarily opposed to our interests—they want our business, after all. But when families show up to policy discussions in schools or engage with local, state, or federal government, it's not an even playing field. With billions of dollars at stake and grand ambitions, Big Tech's already substantial influence will only grow stronger.

The Unprecedented Scale of AI Investment

While informed and engaged consumers have historically been a meaningful counterbalance to industry lobbying, the AI moment feels different in scope and speed. It's not just one company, one tool, or one issue—AI touches every aspect of our lives, from home and work to school. This makes it even more important to maintain healthy skepticism and advocate for protections when it comes to AI policy.

The level of investment in AI supports this bigger story— with companies set to spend more than $300 billion on AI innovation in 2025.[13] When you're investing at that scale, the pressure to show returns is enormous. Companies need to move fast, capture market share, and establish user relationships before their competitors do.

Think about how quickly generative AI companies went from launching consumer applications to having educational partnerships with school districts across the country. That's not an accident—it's a deliberate strategy to embed AI tools into daily family routines (and before policy makers or competitors can catch up). The faster families become dependent on these tools, the harder it becomes to make different choices later.

This doesn't mean we won't benefit from these advancements or that Big Tech won't deliver genuinely helpful learning tools. But the speed at which everything is moving means we often don't have time to thoughtfully evaluate what we're adopting before it becomes deeply embedded in our lives.

Big Money in Industry Lobbying

Tech companies engage with policymakers at every level of government, from city councils considering privacy ordinances to federal agencies developing AI regulations. They hire professional lobbyists, fund think tanks, sponsor conferences, and create industry coalitions that present unified positions on policy issues. This is standard practice in American politics, and in many ways, it provides valuable technical expertise to policymakers who may not understand complex technologies. But it's also an industry that trades in influence and big money —something families and general consumers can't match.

A major AI company might spend more on a single lobbying firm than an entire school district's annual technology budget. They can fund research that supports their preferred policies, bring teams of experts to legislative hearings, and maintain ongoing relationships with policymakers that individual families simply can't sustain. For instance, Meta spends the most of any similar technology company on lobbying the US federal government, having invested \$24 million in 2024.[14]

Amazon came in second, spending more than $19 million on lobbying activities.

This doesn't mean technology companies are manipulating the system, but their voices often dominate policy discussions simply because they're consistently present and well-resourced. While a company can afford to have lobbyists attend every relevant meeting, families are juggling work and childcare responsibilities. The voices at the table are unequal. This is how advocacy works in every industry, but we need to recognize the headwinds it creates when advocating for children's best interests.

Understanding these dynamics helps us engage more effectively in policy conversations, which I'll cover in more detail in Part Three of this book.

Why Children's Data Is Especially Valuable

Tech companies don't just want future users—they want the insights children provide. When it comes to building AI tools, data from children can be particularly valuable for innovation. For instance, researchers at NYU recently trained an AI system using video recordings from a single child from the age of six months to two years old.[15] Using data that represented just one percent of the time the child was awake, the technology successfully learned words and concepts from everyday experiences.

This kind of research could lead to more effective educational tools, better support for children with learning differences, and deeper insights into how young minds develop. But it also creates increased demand for these types of insights and more potential intrusions into our children's lives. Big Tech sees children as revenue-generating opportunities, and families find themselves running interference, always feeling one step behind.

We need practical ways to distinguish between AI applications that genuinely support child development and those that primarily serve commercial interests. As UNICEF highlights in its manifesto calling for better governance of children's data, everyone benefits from better data protection standards.[16] It's possible to protect children's privacy and benefit from data insights without exploiting kids. It just requires better governance, understanding of industry pressures, and a focus on the right policy protections.

Why Schools Are of Particular Interest

Schools represent an attractive market for AI companies, and AI services can be equally appealing to schools interested in innovation. Tech companies have big budgets and can offer new resources that cash-strapped schools struggle to match on their own. The relationship can be beneficial, but it's clearly lopsided in favor of the companies.

From a business perspective, introducing AI systems into schools makes perfect sense. It creates early familiarity with products that students might use throughout their lives while generating valuable data that improves AI systems. But this creates pressure on schools to deploy AI faster than they can properly evaluate the educational value, privacy implications, or whether their educators are ready to use these tools effectively.

Schools don't always have the time or resources to fully scrutinize a potential technology partner's background, motives, or long-term viability. In fact, researchers have found that lower-income school districts are particularly ill-equipped to vet vendors and end up not doing it at all.[17] This raises important questions about whether educational AI use primarily serves learning goals or market development objectives.

Of course, the reality is that it serves both. Companies gain market share and valuable data while schools get tools that may genuinely help students learn. But what matters is that families keep the pressure on to ensure these partnerships ultimately prioritize what's best for kids—not just what's most profitable for companies.

The International Dimension to the Debate

The situation becomes even more complicated when we consider the interests of international technology companies, particularly those based in countries that have different values regarding privacy, democracy, and children's rights.

TikTok offers a clear example of why this matters. The company's terms and conditions allow international partners and their parent company to access global user data. In 2021, TikTok agreed to pay $92 million to settle numerous lawsuits that alleged the app illegally harvested personal data from users and shared the data with third parties, including some in China.[18] Two years later, TikTok's CEO confirmed to the House Committee on Energy and Commerce that US users' data still could be shared with the company's parent company in Beijing, Byte Dance.[19]

From a data privacy perspective, this raises important questions about accountability. Companies operating in the US must comply with US law and be held accountable when they don't adhere to our privacy standards. But when companies are headquartered in countries with fundamentally different legal frameworks, enforcement becomes much more difficult. This doesn't mean these platforms can't offer value, but it does mean families should understand the additional complexity involved in how their data might be accessed and used.

Importance of Advocating for Transparency

The good news is that most US-based companies want to earn trust. They understand that losing the confidence of consumers —especially families—could damage their long-term business prospects.

The path forward means appreciating legitimate business interests while clearly advocating for family priorities. We can support innovation and economic development while insisting on transparency about data use, meaningful parental control, and child-centered design. These goals aren't in opposition to each other.

Transparency matters because it allows families to make informed decisions about AI use while holding companies accountable. When companies are clear about how they collect and use data, what their business models involve, and how they're addressing family concerns, trust can develop even when interests don't perfectly align.

Now that you understand how our information becomes profit and where competing interests might emerge, you can think more clearly about how you should expect to be treated as customers. This whole process works better when the leaders and workers at tech companies represent the same diversity as their customers. Diverse teams are more likely to anticipate family concerns, design with different needs in mind, and build products that work for everyone. This is just one of many reasons why advocating for industry diversity matters so much.

Why Diversity Matters in Data Decisions

When we talk about data as currency, there's one aspect of this ecosystem we often overlook: diversity. Not all data is valued equally, and not all families benefit equally from the AI systems their information helps create. These data systems are built by humans, and it matters that those humans represent the widest range of demographics, values, beliefs, and perspectives.

Why address diversity in a chapter about data economics? Because if we're asking whether any exchange of our data is "fair," it matters who's making the rules about value, fairness, and business priorities. The people building these systems make decisions every day about what data to prioritize, which features to develop, and whose needs matter most. Their backgrounds, experiences, and blind spots shape the AI tools that millions of families will use.

This connects to much larger questions about AI's direction, ethics, and the decisions shaping our future. But it starts with understanding who's currently in charge and what incentives they have when creating these systems.

Diversity's Importance to Problem Solving

Diversity isn't just the right thing to do—there's mathematical reasoning behind it too. In 2007, University of Michigan professor Scott E. Page published research illustrating how diverse groups of thinkers are better able to solve problems than "high-ability" individuals working alone.[20] The logic is straightforward: a group of talented but similar individuals will get stuck on the same problems because they approach them in the same way. A diverse group that works well together can collectively think their way out of challenges that might stump even the most capable individual.

The "homogenization" of talent has always been an issue in Silicon Valley, and not just in racial or gender diversity sense but also in education, life experience, and socioeconomic background. Hiring from a particular profile and background has long been standard practice.

I've seen this firsthand in my experience working in the technology industry. My friend Luanne Calvert recently called out one of the least spoken about issues in a talk she gave for TEDxBerlin—socioeconomic and educational diversity.[21] With decades of experience in consumer technology, Luanne shared that when she was hired at Google, they had to "break" a long-standing rule of hiring only from a pre-approved list of schools. While Luanne didn't attend an elite institution, everything else about her stood out, which is why the company ultimately hired her (and as their creative director no less)—and later even reconsidered this approach, as have many other companies.

Here's what makes this especially relevant to AI development: The technology industry can't build effective AI innovation without diversity, and I don't just mean because consumers are calling for it. Remember how we talked about "scarce" and unique datasets being valuable to AI development? Companies need vast diversity to build superior AI systems. Spotting these opportunities and understanding which datasets matter requires people who innately know what they're looking for because of their own lived experiences. As companies continue to build AI-specific teams, this tension between homogeneous hiring practices and the need for diverse perspectives will only become more apparent.

Understanding the People Building AI Technology

The problem is that the current hiring drive to build AI at top companies reflects the exact "homogenization" scenario Professor Page warned about almost 20 years ago. The technology industry remains overwhelmingly dominated by young, white, male engineers and executives with advanced degrees from elite institutions.[22] And it may end up slowing down AI innovation eventually.

Right now, we're seeing news of multimillion-dollar compensation packages and fierce competition between Big Tech companies for talent. But when products start to miss expectations or disappoint consumers, the calculus could shift quickly. Building AI systems requires specific technical skills, yes, but it's not all math and science. Built into those systems are philosophical rules and ethical guardrails that require diverse thinking. They also require diverse data, which becomes harder to obtain when companies lose consumer trust.

When AI development teams lack diversity, they tend to focus on challenges that impact affluent, educated communities while overlooking issues that matter to families from different backgrounds. They might create tools that benefit adults without considering the impact on kids or ignore ethics and other "softer" but highly critical areas of work. Diversity in AI development isn't a "nice-to-have"—it's essential to making any of this work in practice.

Problem of VCs Funding Their Reflection

The lack of diversity in AI development also creates a ripple effect in what types of AI businesses get funded and developed. Investors tend to fund technology leaders that look like them— often white, male, and well-connected. According to research from Harvard University, all-female founding teams receive just 2.4% of venture capital funding, and this rate has remained stubbornly steady over the years.[23] Additional research from Columbia University finds a similarly picture when looking at racial diversity, with just 3.47% of Black founders receiving venture funding.[24]

The problem is typically attributed to something called "homophily," which is the tendency to be most interested in those like yourself. This creates a vicious cycle that can also spill over into AI output.[25] This can impact everything from the type of technology that gets created to what societal concerns we tackle. If we don't pay attention, the world of AI will reflect the interests of just a very narrow group of individuals—and that will end up hurting industry as much as consumers.

The Data Diversity Gap Problem

Most AI datasets overrepresent certain populations while underrepresenting others. This happens because it's easier and cheaper to collect data from populations that are already digitally engaged, and many online communities and platforms share similar demographic profiles among their users. Addressing this requires diverse developers and leaders who can spot the gaps.

A lack of diversity in data hurts everyone, including the companies creating AI systems. Voice recognition systems struggle with different accents. Image recognition systems perform poorly on people with darker skin tones. Educational AI misses learning patterns common in certain cultural communities. This represents both a failure for consumers and an economic loss for the industry.

Adjusting for data bias isn't easy, and attempts to correct it can introduce new problems. We've already seen this happen in embarrassing results shared in the media from AI systems "over-correcting" for bias and producing historically inaccurate images or information.[26] And this is precisely why we need diverse teams to come together and take on these challenges. Diversity in AI development isn't just the right thing to do anymore—it's the only approach that will actually work.

Impact to Big Tech's Bottom Line

Year after year we discuss the importance of diversity, yet little seems to change in the technology industry. Even as the concept of diversity becomes politically fraught, with different presidential administrations taking varying approaches, there's been minimal movement to meaningfully address the imbalance in Silicon Valley. A different approach is clearly required.

Here's the good news: nothing shapes policy or forces follow-through on promises more than an impact on a company's bottom line. And AI has fundamentally changed consumers' power to collectively advocate for change. Companies building AI systems need our data—without it, building, training, and advancing AI innovation isn't possible. Meanwhile, billions of dollars are being invested in AI with little return so far.

According to a recent McKinsey & Company report, nearly eight in 10 companies are using AI, but the vast majority have seen no meaningful financial impact.[27] Companies need us. They need our data, and they need us to use their technology. They need a diverse user base. This gives us real leverage to demand that the people creating these technologies better represent society. Now is the time to push for this change and to prepare our kids and others to join their ranks.

Key Chapter Takeaways

After everything we've covered in this chapter, you might be feeling equal parts hopeful and concerned. We've explored some significant realities about how the digital economy actually works and where families fit into this ecosystem. But I hope the bigger takeaway here is our collective economic power.

This empowerment comes from recognizing that your data isn't just meaningless information—it's incredibly valuable and in-demand currency. Your data is important, and you own it. The awareness you've built in this chapter transforms how you approach future digital decisions. When companies introduce new features, change privacy policies, or launch new platforms, you now have frameworks for evaluating whether these changes serve your needs or primarily benefit corporate data collection goals.

What excites me most about working with families on these issues is watching adults become empowered advocates for better digital practices. Your individual choices matter, and they're part of a larger movement toward more equitable digital experiences. When families collectively demand fair value for their data, the entire industry responds with more privacy-respectful options.

Shifting Our Thinking About Data

I hope you now see "free" apps, platforms, and services in a new light. These digital tools operate on hidden economic exchanges where your data generates ongoing revenue streams that often far exceed any direct benefits you receive. They're not free to us and arguably never have been. The only difference is that we thought digital advertising was the exchange, not data collection.

When your child downloads an educational app or you sign up for a new platform, you're not just getting convenient tools—you're undertaking a financial transaction. Understanding this doesn't mean avoiding these services entirely. It means approaching them as an informed consumer who knows what you're trading and whether you're getting fair value in return.

We also talked in this capter about how companies treat you differently depending on whether you're the customer or the product. When you're the customer, companies prioritize your satisfaction and make it easier for you to control the experience. When you're the product being sold to advertisers, companies optimize for data collection and engagement rather than your family's well-being.

You now see how important it is not to resign ourselves to this data collection fate. As Professor Zuboff says in *The Age of Surveillance Capitalism*, "We succumb to the drumbeat of inevitability, but nothing here is inevitable. Astonishment is lost but can be found again."[28] Let's stay astonished about what's happening and work to change it.

Diversity Factor in Making Good Data Decisions

We've also explored how the lack of diversity in AI development affects the value and fairness of data exchanges. When the people building AI systems come from similar backgrounds, they naturally prioritize solving problems they understand while overlooking issues that matter to different families and communities.

This awareness helps you ask better questions. Does this educational tool work well for children like mine? Was this AI system trained on data that represents our family's experiences? Are we supporting companies that value diverse perspectives in their development processes?

Understanding these dynamics doesn't mean avoiding all AI tools. It means approaching them with informed skepticism and supporting companies that demonstrate genuine commitment to inclusive development.

Looking Ahead to Stronger Data Ownership

This might sound like a lot to keep track of for every digital decision, but remember that understanding your data as currency means making informed, strategic choices that serve your family's long-term interests while still allowing you to benefit from these digital tools and opportunities.

Now that you understand the economic dynamics at play, we need to address how commercial incentives can cause emotional harm. The companies that profit from your family's data don't just collect information—they use sophisticated techniques designed to capture and hold your children's attention. The same AI systems we've discussed can create powerful psychological dependencies that impact how kids think, feel, and relate to others.

Understanding the economic motivations behind these platforms prepares you to recognize when engagement strategies cross the line from helpful to manipulative. When you know that a company's revenue depends on keeping your teenager scrolling, liking, and sharing, you can better evaluate whether their platform design truly serves your child's best interests.

In the next chapter, we'll explore how the profit motive can result in psychological harm, whether intentional or not. Armed with your new understanding of data economics, you're in a better position to address potential issues before they arise.

You're no longer passive consumers in this newly amplified digital economy. You're informed participants who understand the true value of what you're trading and can make choices that serve your family's best interests. That knowledge is power, and it's precisely what you need to protect your children in the years ahead.

Chapter 6
How AI Can Challenge Kids' Mental Health

Recently, my friend Lisa shared a worry she had about her 15-year-old daughter. "My daughter used to come home from school bursting with stories about her friends and teachers," Lisa said. "Now she barely mentions anything about her day, but I know she's having long conversations with ChatGPT about everything from friendship drama to her future career plans. It's like she's replaced human connection with AI interaction, and I don't know if I should be worried or grateful that she's sharing her feelings"

Lisa's experience reflects one of the most nuanced challenges families face with AI. The risks aren't always obvious like deepfakes or fraud. Sometimes they're as subtle as a child preferring an AI chatbot conversation to a human one. This creates a new kind of gray area where AI may be as helpful for social and emotional development as it is harmful.

Watching a child form a "relationship" with a machine can be unsettling, especially when it seems to come at the expense of connections with family or friends. Yet it's difficult to reconcile this concern with the fact that a chatbot may provide genuine comfort and support for a child. With the AI industry expanding at breakneck speed, this tension isn't going away.

As with much of what we've explored in this book, addressing these challenges starts with awareness, advocacy, and staying true to your family values. It also requires listening to our children, understanding their perspectives, and encouraging healthy skepticism.

Many voices are already demanding that AI companies, legislators, and government bodies prioritize oversight of AI chatbots and their impact on kids. Staying informed about changes as they happen and adding your voice to these conversations matters.

Addressing AI "Companionship"

The idea that we must help kids navigate relationships with other kids AND machines seems absurd. Even more challenging is that there isn't just "one kind" of AI system to consider from a social and emotional health perspective. There are mental health chatbots designed for therapeutic support, general platforms like ChatGPT, chatbots integrated into social media platforms and online communities, and a new category of AI systems intended solely to be human "companions." We'll explore these more thoroughly in the pages ahead, but companion bots can be the most concerning when it comes to their impact on isolation, social development, and dependency.

For children struggling with anxiety, depression, social challenges, or family stress, AI conversations may be genuinely helpful and shouldn't be dismissed entirely. Harvard researchers have found that AI chatbots can help reduce feelings of loneliness, with many users reporting that "feeling heard" was one of the most positive aspects of their interactions with these systems.[1]

What makes navigating all of these different systems tricky for families is that the mental health implications often develop slowly. Unlike more visible concerns like inappropriate content, challenges to social and emotional development can be subtle and easy to miss until they become serious. Companion chatbots can also take many forms, showing up integrated into social media platforms, games, and niche online communities where their influence is harder to detect.

How AI Challenges Truth and Reality

AI-driven social and emotional challenges don't stop at companionship. Kids also have to navigate synthetically created fake content that ranges from confusing to weaponized. Even the possibility of becoming ensnared in a deepfake situation can cause enormous anxiety.

According to child-protection nonprofit Thorn, incidents of deepfake "sextortion" are rising with alarming speed. This happens when a compromising image—often fake—is used to threaten or extort money from a child. Thorn's survey found that 20% of kids ages 13-20 have experienced some form of image-based extortion.[2]

AI technology is also amplifying the social and emotional issues families have been dealing with on social media platforms for years. On the backend, AI can supercharge the entire social media ecosystem. It can more precisely personalize user feeds, optimize targeted advertising, and automate content creation. But it can also amplify bias, create more privacy concerns through increased data collection, and serve as a vehicle for sharing harmful and targeted deepfakes.

AI-generated "bot" accounts can also damage trust, as they might appear as real people that are vehicles for manipulation and disinformation. Unfortunately, researchers at the University of Notre Dame have found that protections against these malicious automated accounts have been woefully inadequate.[3] These industry shifts are difficult to keep up with, and their impact can be profound.

What You Will Discover in This Chapter

In this chapter, we'll explore how AI is reshaping and testing children's social and emotional development. You'll learn to recognize when AI crosses the line from helpful emotional support to replacing human relationships. We'll examine troubling developments like deepfakes that can alter reality and destroy trust. We'll also tackle how AI amplifies the mental health risks kids already face online and look at what AI "companionship" may mean for children's ability to navigate real-world relationships.

There are so many things that AI systems can't replicate—the productive conflict in relationship development, the reading of emotional cues, compromise, forgiveness, and trust building. AI relationships, no matter how sophisticated, can't capture the beautiful complexity of human social dynamics, and children shouldn't be robbed of this experience.

We can also help kids navigate the emotional pressures that AI brings by building skills related to personal agency and independence. When children feel more confident, think for themselves, apply critical thinking skills, and don't worry about being "different," they build the necessary emotional resilience to protect themselves.

The truth is that regardless of how we feel about AI innovation and its opportunities and risks, we must teach our children to exist harmoniously with these new systems. Strategies for building stronger social and emotional skills will help all of us both survive and thrive in the years ahead.

AI's Impact on Social and Emotional Development

Media coverage of AI and kids focuses heavily on the risks generative AI poses to education, but we hear far less about the challenges to children's social development and emotional growth. By inviting "human-like" technology into our kids' lives, we've introduced a potential obstacle to their development in ways we should be discussing in much more detail. Cambridge University Sociologist Nomisha Kurian has noted as much in her work.[4] She writes that AI's ability to "simulate" human connection can have an impact beyond whether these tools are useful for learning and should be given far more consideration by adults.

There's no handbook for supporting kids in developing social and emotional skills even in the best of times. Now we're dealing with technology that can amplify existing challenges, like social media, while also introducing new variables that test kids' ability to separate human interactions from those of machines. All of this is happening during a time of life that's already emotionally challenging to navigate.

The picture gets more complicated because AI chatbots may actually have a genuine place in supporting kids' mental health and social development. But the boundaries remain unclear, and families have few resources to figure it all out. Our children are digital natives who adapt faster than we do to these new changes and challenges, so getting their feedback is important too.

Risk of Bypassing Skills Development

Social skills develop in ways that are often spontaneous and unpredictable. Kids learn to read emotional cues, navigate conflict, practice empathy, and build trust through countless interactions. Children also learn through the hardest and most emotional experiences—the pain of rejection, disagreements with siblings, the ups and downs of complex social dynamics at school.

There's no meaningful replacement for these in-person interactions. Skills like reading non-verbal cues and emotional subtleties, navigating uncertainty, practicing patience, and building trust are acquired through small and seemingly insignificant moments that happen each day. Human relationships teach children that social interactions involve genuine reciprocity, emotional risk, and the complexity of managing multiple perspectives at once. This is part of growing up.

When AI systems interfere with this process, the impact can be lasting. We need to pay closer attention to how any sort of "AI companion" serves as an insufficient substitute for doing the work of relating to one another. Recent research found that the more satisfied adult users were with their AI companion interactions, the more their real-world relationships suffered.[5] If this is true for adults with fully developed social skills, the implications for children who are still learning these fundamental abilities deserve serious consideration.

The Lure of the Sycophant Machine

It's concerning enough when adults retreat into AI interactions, but for children who need to practice social skills, AI can be a poor substitute. It's also one that tells us exclusively what we want to hear, all the time. Much has been written about the "sycophantic" nature of generative AI chatbots, particularly with the initial release of OpenAI's ChatGPT version GPT-4o.[6] What's most worrisome about this machine-based "agree-ability" is how it can reinforce anger, delusion, or other negative tendencies in a fragile person interacting with the system. OpenAI quickly addressed the issue, but it highlighted the risks of such overly positive reinforcement—and how easily we can be mesmerized by sappy encouragement.

The biggest issue with the always chipper and complimentary tone of AI chatbots is how they can encourage kids to avoid the hard stuff that's critical to gaining maturity. From understanding awkward silences, disagreements, and hurt feelings to learning forgiveness, empathy, and compromise, AI interactions can reduce children's ability to navigate the inevitable discomfort of human relationships. They can also encourage unequal and even pathological "take" versus "give" behavior.

Consider a child who becomes attached to an AI system's perceived flexibility but then struggles when friends aren't as available. The predictable, patient responses from AI don't prepare children for the reality that human relationships require accepting that others won't always respond as expected. Kids need practice navigating these disappointments and adjustments—it's how they learn resilience and adaptability.

AI-Influenced of Life Decisions

A neighbor who works as a guidance counselor told me she's noticed kids increasingly using AI to make big life decisions, including asking for feedback on what college to choose and what career path might be "best" for them. "These kids are having deep, personal conversations with AI systems about their futures," she explained. "The AI chatbot responds with such confidence and seeming understanding that students trust it more than human advisors who actually know them."

What my neighbor identified is how AI systems can facilitate a form of social engineering. AI systems apply statistical patterns to personal decisions, nudging kids in one direction or another based on demographic assumptions drawn from their data. This can run the gamut, from steering kids to certain colleges to influencing shopping and entertainment purchases to even guiding political and social beliefs by creating echo chambers. Rather than truly understanding a child's interests, aptitudes, and capacity to break from expected patterns, an AI system can lead them down a "statistically prescribed" path.

You might think, "How is asking ChatGPT for a list of colleges any different from searching on Google?" The difference lies in how AI presents the information. Google provides search results that children understand come from various sources they need to evaluate. AI delivers personalized advice that feels like trusted guidance rather than algorithmic predictions. It makes kids feel that an AI chatbot "gets them" but ultimately discourages them from gathering different perspectives or thinking for themselves.

Amplifying "Echo Chambers"

Beyond understanding how AI systems can inadvertently "engineer" a child's choices, families should also watch for signs that a child might have fallen into an "echo chamber." This happens when AI doesn't just influence individual decisions but shapes a child's entire worldview through sustained, personalized algorithmic interactions.

Cambridge University academic C. Thi Nguyen wrote about this issue in 2018.[7] In his paper "Echo Chambers and Epistemic Bubbles," he notes that fundamentally echo chambers work by manipulating trust. When it comes to approachable, authoritative chatbots, we need to ensure that trust stays with adults and not with machines.

Social media platforms already tend to show users content that confirms their existing beliefs and interests, which can significantly affect developing identities and worldviews. AI amplifies this effect by generating unlimited amounts of content tailored to reinforce whatever perspectives keep an individual user most engaged. Instead of just curating existing posts and comments, AI can create new interactions that feel authentic but are specifically designed to validate and intensify particular viewpoints.

The conversational nature of AI means that influence doesn't rely on static posts but can engage children in ongoing dialogues that adapt based on their responses. This makes manipulation feel like natural conversation rather than targeted content. AI systems can also insert chatbots to further conversations and themes or create false urgency around decisions.

For children still developing critical thinking skills and forming their identities, recognizing these AI-amplified echo chambers becomes an important part of digital literacy. Teaching kids to seek diverse perspectives and question sources

—even friendly, helpful-seeming AI—helps them develop the discernment they'll need.

Finding the Right Balance

If used thoughtfully and in moderation, AI chats can be a useful tool for working through some of life's challenges, as the Harvard research I mentioned noted. But this is for each family to consider and decide individually, in line with their beliefs and values.

Families can ensure that AI social support supplements rather than replaces human relationships by staying actively involved in discussions with kids about their social and emotional development. We can also model good behavior through our own personal interactions and how we as a family socialize and build relationships together.

Establishing clear family expectations that important life decisions require human consultation is essential. AI can help kids think through options, but the final choices about relationships, education, and personal values must involve real people who know and care about them. It's also helpful to remind kids that we're learning too and want to know what they're finding when they interact with AI systems.

This foundation of trust and open communication becomes even more important when families face some of the more serious risks we'll explore next, such as deepfakes that can disrupt a child's fundamental sense of what is real. By deepening communication with our kids now, we can be much more effective in supporting them as we all navigate these new challenges together.

Shannon Kimberly Edwards

When Deepfakes Destroy Trust in Reality

No doubt most families have become familiar with the AI-driven new world of "deepfakes." As I covered in Part One, these are artificially created, or "synthetic," audio, imagery, or video content, sometimes created for fun but often to deceive or even extort users. They are also an equal-opportunity threat, having touched everyone from unsuspecting suburban teens to celebrities.

When a sexually explicit "deepfake" image of Taylor Swift was shared for just one day in January 2024 on X (formerly Twitter) and via 4chan, it amassed millions of views.[8] Although it was taken down less than 24 hours later, the image had already become news, and X was forced to disable any search for the singer until the controversy subsided. It was a dramatic turn that if you blinked, you might have missed, but the impact was lasting.

Because at the same time as Swift's high-profile experience, New Jersey high school student Francesca Mani and her family had already begun fighting back after a similar personal violation. In 2023, Mani learned that a group of boys in her class had used AI software to fabricate sexually explicit images of her and other female classmates. It was one of the earliest documented uses of AI in this way, and Mani had the courage to take her case to legislators. In 2024, *Time* named her one of their *100 Most Influential People in AI*.[9]

Unfortunately, the use of AI technology to create synthetic media to defame, defraud, humiliate, and bully adults and children has become one of the most serious AI-driven threats that families face right now. According to the European Union, an estimated 8 million deepfakes will be shared by the end of 2025, compared to just 500,000 shared in 2023.[10] Even worse, they note that pornographic material accounts for 98% of this "synthetic" content. The issue is worsened by a lack of education about the dangers of this trend with families, kids, or educators, and how it can have an impact that follows young people into adulthood.

What makes deepfakes particularly challenging for families is how the technology weaponizes our natural trust in visual evidence. We're wired to believe what we see, yet AI has advanced to the point where synthetic media can fool not just casual observers but sometimes even experts.

When children regularly encounter realistic fake content, it can undermine their fundamental trust in what they see and hear. This erosion of trust affects kids' ability to make informed decisions, maintain healthy relationships, and engage constructively with information throughout their lives. Understanding what's happening is the first step families, educators, and legislators can take to help protect kids from this content.

How Deepfake Technology Works

Like the other AI technologies we've covered in earlier chapters, targeted deepfakes result from machine learning algorithms that analyze thousands of images or videos of a person, mapping their facial expressions, head movements, and speaking patterns. The AI then applies this learned information to substitute the target's face and voice onto different video content, creating realistic but completely fabricated footage.

In more technical terms, deepfakes result from what is called a "generative adversarial network," where two AI systems essentially compete against each other.[11] One creates fake content while the other tries to detect it. Through this process, the fake content becomes increasingly convincing until even the detection system can't tell the difference.

Modern deepfake tools have become so user-friendly that

almost anyone can create this type of media. Some apps can swap faces in real-time during video calls, while others can generate convincing audio clips from just a few minutes of someone's voice recording. The quality varies, but even lower-quality deepfakes can be effective for harassment or extortion purposes. While major platforms have pledged to address the issue, enforcement remains inconsistent and and those determined to create harmful content continue to find workarounds.[12]

For families, this means that any photos or videos your children share online—from Instagram posts to TikTok videos to school yearbook photos—could potentially become source material for deepfake creation. The abundance of visual content that teenagers naturally share makes them particularly vulnerable targets.

Terrifying "Kidnapping" Scam

Another deeply disturbing application of deepfake technology is voice cloning for fraudulent purposes. The technology has advanced to the point where AI can generate realistic speech using just a few minutes of recorded audio. For kids who regularly post videos on social media or participate in virtual class discussions, enough voice data may already be publicly available to create persuasive audio deepfakes.

Arizona mother Jennifer DeStefano shared her harrowing personal story with Congress in 2024.[13] She received a phone call from someone she believed was her daughter, saying she had been kidnapped and that the kidnappers were demanding ransom. The call was so realistic that DeStefano was left shaken. Even after speaking to her husband and confirming that her daughter was safe, DeStefano struggled to reconcile the facts with what she had heard.

Audio-based deepfake crime has become such a threat that US government agencies have even been forced to issue consumer warnings.[14] What should you do if this happens to your family? Officials suggest that if you receive a call claiming a family member has been kidnapped, stay calm and ask to speak directly to the loved one. Ask for details that only your family member would know (such as a family "safe" word). While staying on the line, use another phone to contact the same family member to verify they're safe. Then call the police. Most importantly, trust your instincts—if something isn't adding up, you're likely right. As this becomes more prevalent, law enforcement is better prepared to address it, so contact them as soon as you can.

Real Consequences for Kids

The scope of potential harm ranges from immediate social consequences to long-term impacts on kids' futures. Malicious actors can exploit children's natural curiosity about new technology, turning experimentation into opportunities for bullying and harassment. Deepfakes can be used to create false evidence of academic misconduct, behavioral problems, or rule violations. Kids might face suspension, disciplinary action, or academic consequences based on completely fabricated evidence that appears authentic to school administrators and teachers.

Fake content can also become part of children's permanent digital footprints, potentially influencing college admissions, scholarship opportunities, employment prospects, and personal relationships years into the future. Even when content is eventually proven fake, the initial impression and associated controversy may continue to influence decisions about a child's future.

Exposure to deepfakes, whether as targets or observers, can

have profound psychological effects on children.[15] Regular exposure to realistic fake content can undermine kids' confidence in their ability to distinguish truth from falsehood.

This uncertainty can impact their decision-making, critical thinking development, and fundamental trust in evidence and authority. When children discover that realistic fake content can be created by anyone, it may damage their ability to trust authentic communication and relationships. They might become overly suspicious of legitimate content or, conversely, become more vulnerable to actual deception.

Every Step We Take Helps

Every step a family takes, however small, makes a difference. Start by reviewing the photos and video content of your children that may be in public accounts and consider what can be removed. Even just limiting the photos we share of our kids online can offer meaningful protection. While social media platforms are developing detection tools and policies to combat deepfakes, these measures remain imperfect right now, though they're evolving rapidly.

The most effective protection comes from having honest discussions with our kids about these risks. Families should also establish protocols for verifying unusual communications and preparing for potential extortion scenarios. Each of these conversations is an opportunity to connect with our kids and hear what they have to say. This generation is more digitally savvy and aware of technology like deepfakes than any before. We want to keep these conversations going so they don't get lured into digital rabbit holes—including the world of AI companions, which we'll explore next.

New World of AI Companions

In a remarkably short period, we've settled into a world with decreased human interaction. The convenience of the Internet and synthetic connection that social media brings have now been joined by AI chatbots simulating humanity. The question is whether these more recent advancements, such as AI "companions," represent a step toward digitally-driven isolation or a way out of an epidemic of loneliness.

We're not just talking about everyday generative AI chatbots anymore, but completely customizable chatbot "friends" that learn about users with every interaction. These purpose-built companions aim to be the "perfect" friend, romantic partner, or whatever role the user desires. They can present as realistic avatars, and users can select gender, hairstyle, eye color, and clothing. Some apps even have "personality sliders" that allow users to adjust traits like shyness or flirtatiousness. Their interests are customizable, as is the character's "backstory." It's a fantasy world where a user is the main character and the companion exists to fulfill whatever emotional needs they have.

AI May Help With Loneliness

In 2023, the US Surgeon General released a report on the "epidemic of loneliness" in the country, noting that the health risks of being socially disconnected are similar to those caused by heavy smoking.[16] For some adults—seniors or those with mobility or other challenges that keep them homebound—AI "companionship" may lift spirits and provide connection. We shouldn't rule out the usefulness of these systems without considering groups that struggle with connection, though this must be done thoughtfully.

For kids, there hasn't been enough research or education

about the psychological risks that AI companions present. Many families have no idea these companionship companies even exist. According to a report by the Australian Government's eSafety Commissioner, by early 2025, there were more than 100 of these AI businesses, including familiar names such as Character.ai and Replika.[17]

The FTC Steps in to Get Answers

Companions are also widely available on social media networks, which may come as a shock to families. Meta, in fact, launched its "create your own AI" feature in 2024.[18] Today their marketing text announces that not only can you create a companion, but "creators can even build an AI extension of themselves"—a simulation that can chat across the Meta network as if the AI system were you. Kids can now create AI companions on their favorite social media platform and talk to bots created by other users across various platforms. Children might think they're chatting with a friend when it's actually the friend's bot, or even a stranger pretending to be a kid.

In September 2025, the FTC launched an investigation into AI technologies used for this very purpose.[19] The Commission has asked companies to provide details about which age groups are using these customized chatbots, how much time kids and teens are spending with them, and which AI characters are most popular with the youngest users. They're also investigating how often minors are engaging in inappropriate conversations with these AI companions—something Meta was accused of allowing for in a leaked internal document.[20]

The FTC is also demanding transparency about safety measures. Do these companies have age restrictions? Are there parental controls available? How do they handle children's personal data differently from adult users? What steps are they taking to identify and prevent potential harm to young users?

This government investigation signals that regulators are finally taking seriously the mental health and safety risks we've been discussing. But families need to understand the scope and scale of this landscape because it's not going away, even with better regulation.

Kids Are Moving Faster Than We Can Keep Up

Kids are adopting this companion technology far faster than adults can catch up—or regulators can try to rein the industry in. According to a 2025 study by Common Sense Media, 72% of 13-17 year olds have used an AI companion at least once, while 52% said that they interact with these platforms at least a few times a month and 13% do daily.[21]

This is a global phenomenon. Chinese cultural newspaper *Jing Daily* recently wrote about the app "Soul," which has become incredibly popular with Generation Z in Asia—seeing its engagement numbers increase 76% year-over-year.[22] The paper notes a survey by Soul finding that 70% of respondents turn first to their companion to share feelings, and nearly 33% said they prioritized an AI companion's connections over romantic relationships.

In many ways, while adults have been debating generative AI in schools, kids have moved on to making machine-generated friends. Focusing on both is both possible and necessary, as it will be difficult to redirect kids away from these relationships once they're established.

Engineered Emotional Bonds

These engineered emotional bonds are what should concern us the most. Researchers at MIT have called this new category of AI "addictive intelligence."[23] They note both the highly experimental nature of these relationships and that users might not fully understand what they're getting into when they sign up. There is no "consent" button for an addictive digital relationship.

When kids regularly experience AI's simulated empathy, 24/7 availability, and consistently supportive tone, they may more easily reject the emotional limits and reciprocity requirements of human relationships. Relying on AI companions for "love" and "support" may make it difficult for individuals to accept the imperfections and challenges of a real-life partner.

Developing genuine empathy requires hard work. It means learning how to apologize, rebuild trust after conflicts, and maintain relationships through difficult periods that call for forgiveness and growth. A chatbot companion offers none of these challenges or opportunities to develop crucial emotional skills.

Practical Considerations for Families

One of the most effective approaches is making sure kids regularly interact with family members, friends, and community members without AI assistance or digital mediation. While it might seem obvious, in-person interaction is now more important than ever.

Daily family conversations without devices, face-to-face social activities, community service involving direct human interaction, and family traditions that require emotional presence all help children experience genuine emotional exchange, conflict resolution, and relationship building.

We also need to model in-person connection and commitment to cultivating relationships. Simple actions like sharing our feelings and daily experiences, showing how to comfort others even when we're not feeling our best, and expressing authentic appreciation demonstrate these values.

We need to help kids understand that human emotions are complex, changeable, and sometimes difficult to navigate—but that this complexity represents what makes us human. Part of this journey is teaching kids to advocate for these things themselves, which brings us to the importance of personal agency, autonomy, and individuality—subjects we'll explore next.

Personal Agency, Autonomy, and Resilience

When considering kids' social and emotional well-being in the age of AI, the final issue we'll explore is how to protect children's power to think for themselves. This doesn't just mean the difference between letting ChatGPT write a paper on World War II versus doing the work themselves. It means ensuring kids can resist coercion from AI-driven social media, reject the flattery of sycophantic AI, and question synthetic content despite what they see with their own eyes.

It's about standing firm and thinking independently—one of the most important skills kids will need for the future ahead. Letting AI systems make decisions while avoiding the hard work of human interaction creates a troubling combination. We need strong, critically aware leaders in the years ahead.

Why Personal Agency Matters

There's been far more discussion about the educational benefits of AI than how the technology may negatively impact a child's mental health, particularly their ability to make their own choices and decisions. Researchers have started to note in their work—specifically that AI assistance does appear to have the potential to discourage personal agency.[24] A child without the ability to confidently make their own decisions is even more vulnerable to AI's emotional influence.

A sense of control is central to a child's emotional well-being. When children feel they have the power to make meaningful choices about their lives and the technology they use, they build confidence for even bigger and more challenging decisions. Conversely, kids who are overly dependent on AI for decision-making—from what to wear to how to respond to friends—risk losing that critical capacity.

The Oxford University Child-Centred AI Design Lab has made individual children's "agency" core to their work.[25] The researchers believe that a technology design approach centered on kids' power to make good choices and identify potential harms themselves is more effective than restricting access to technology. This means including kids' feedback in the design process and considering them as users in every step of product development.

This approach recognizes what many families instinctively know—our children, as digital natives, often understand technology in ways we don't. We should give them the chance to lead and practice being in charge.

Dramatic coverage of AI as an existential threat, suggesting we'll lose control as a society to robots, can also chip away at children's sense of personal agency. Continuing to tell tomorrow's leaders that they won't have any control may make these dystopian scenarios self-fulfilling prophecies. What we tell children today will influence who they think they can become tomorrow. It's in our best interests to empower the youngest generations and remind them that machines are nothing without us—and that it's today's kids who will lead us into the future.

Building Emotional Resilience

Psychologists increasingly consider positive mental health to be marked, in part, by a person's level of "emotional resilience"— one's ability to positively adjust when faced with stress, trauma, or adversity. Research suggests that while much can innately influence emotional resilience, such as a person's biology, we can also learn these mental health strengthening skills.[26] It's something we'll explore more in later chapters, but as a response to any challenges posed by AI, it's a powerful one.

Licensed clinical social worker Karen Herbert, who helps build emotional resilience in her work focused on women and families, is a lifelong friend of mine.[27] She reminds us that emotional resilience has always been important and is strengthened via the same good parenting skills that existed in our pre-digital world.

"It's important to help children focus on what they can control and remind them that anything that feels bad won't persist forever," Karen said. This can mean actions as simple as telling kids that you're interested in how they feel and that they can rely on you for support. Sometimes we forget how powerful honest and emotionally rich conversations can be for helping kids face challenges.

We should encourage children to explain their feelings, support their arguments with facts, and then give them the space to learn from any consequences. But in all cases, the key is letting them know that we're here as a safety net.

One of the hallmarks of the years ahead will be peaks and valleys of stressful disruption. Building emotional resilience can help us weather it all and will increasingly be the type of skill we look for in our future leaders.

Creating Emotionally-Healthy Future Leaders

Just like we prepare kids to leave home, we need to get them ready to make their own decisions around AI and to have the strength to resist issues such as coercion or groupthink. While we don't know what AI capabilities will exist when our children are adults or what psychological pressures they'll face, we do know the type of skills that make a good leader. We can prepare by encouraging self-awareness and critical thinking skills and instilling the confidence they'll need to protect their mental health and strengthen their personal agency.

The more than 800-page *Oxford Handbook of Ethics of AI* was published in 2020 and serves as an illustration of how much thought has gone into AI and ethics over the years.[28] But one of the simpler points made by the academic authors is this: by definition, "intelligence" means the ability to make the right decision at the right time and the ability to consider context.

This matters because it means AI doesn't signal some dystopian future where robots outsmart us. Instead, it reminds us that we possess the innate intelligence that gives us power, control, and leadership—and we're simply teaching machines to handle some of what we already know how to do. The contextual judgment that defines true intelligence remains fundamentally human.

The truth is, when it comes to AI innovation, it's not bigger than us. We can decide to give up control or to use this innovation opportunity to make our lives richer and better. Just as we slowly get our kids ready for other milestones in life, we should let them learn to take the lead in making thoughtful choices about AI integration that supports their wellbeing and growth.

Key Chapter Takeaways

Since the launch of ChatGPT, families have been thrust into a complex new world of technological innovation with profound implications for our children's mental health and social development. Understanding AI's opportunities and challenges touches every aspect of our lives and has become as much of a mental health consideration for adults as it is for kids. But as families explore this technology together, we can build resilience, embrace our control, and navigate the future as partners. There are hard issues to consider, but in every case, having honest, open discussions moves us forward.

There's never been a better time to share our own personal vulnerabilities and feelings of uncertainty with our kids. AI is turning all of our lives upside down and making us face enormous questions about our humanity. That's a lot for families to wrestle with. AI is coming at us from all directions, from systems that "mimic" humans to technology that amplifies the platforms and tools we already know.

What I hope you take away from this chapter is that AI can simultaneously support and threaten healthy development. AI systems are just that—systems and technology, not new overlords that demand our loyalty. We are in control. Finding the right balance between the risks and opportunities is ultimately up to all of us. You get to decide what's best for your family and can flip the switch to "off." Your family values, comfort with communication, awareness of pitfalls, and intentional choices will make the path forward clear.

Impact of AI on Social and Emotional Growth

Perhaps the most crucial issue we need to understand is that AI's effects on children's social and emotional development are rarely obvious or immediate. Unlike traditional concerns like cyberbullying that produce visible consequences, the challenges we've explored—from bypassing essential social skill development to becoming overly dependent on AI for emotional support—typically develop gradually through seemingly positive interactions.

AI chatbots and companions provide simulations of support and understanding that can feel genuinely helpful while subtly undermining children's development of essential human skills. This gradual nature means that early awareness and honest conversations are far more effective than waiting until problems become entrenched.

Protecting Our Trust in Reality

We've seen how deepfakes and synthetic media are threatening not just our children's safety but their fundamental trust in what they see and hear. The psychological impact extends beyond immediate targets, potentially undermining children's confidence in authentic communication and their sense of what's real. This isn't just about teaching kids to spot fake content—it's about helping them stay grounded in reality and ready to defend what's real.

The rise of AI companions deserves particular attention. Designed to create emotional bonds for commercial gain, these tools should be at the top of families' watch lists. When children become accustomed to AI's consistently supportive, always-available nature, they may struggle with human relationships that involve natural limits, emotional complexity, and the requirement of genuine reciprocity. New interventions from the federal government may help slow the popularity of these tools with kids, but family awareness remains crucial.

Preserving Personal Agency

The mental health implications we've discussed all point back to the critical importance of personal agency. When children can make thoughtful decisions about AI use, reflect on how these tools affect their wellbeing, and advocate for their needs, they're building the emotional resilience they'll need throughout their lives.

This means helping them recognize when AI use supports their growth versus when it becomes a substitute for developing their own capabilities. It means encouraging them to maintain human connections that provide the emotional give-and-take that AI cannot replicate. And it means celebrating kids' independence when they choose the harder path of human interaction over the easier route of AI assistance.

Your Family's Personal Approach

The way your family approaches these challenges should be unique to you, based on your individual values and your children's needs. What matters most is approaching these decisions intentionally rather than passively accepting whatever technology companies provide.

The social and emotional skills our children build now—through messy human relationships, navigating real conflict, and maintaining their sense of personal agency—will serve as the foundation for thriving in an increasingly AI-integrated world. We should stay flexible as our children's needs and AI capabilities continue to evolve, but also keep our focus on the fundamental human development needs that remain constant.

It's about preparing kids not just to use AI systems effectively but to maintain their essential humanity while benefiting from technological advancement. When families work together with clear communication and shared values, children can develop both digital literacy and the irreplaceable human skills that will serve them throughout their lives.

You hold real power here, and you have control over the future. This is technology built by humans, and as such, we have a say in where it all goes. Hold tight to this message as we head into the last part of the book, where we'll celebrate the ways we can more actively use our power to shape the future.

Part Three
The Path Forward: Thriving in an AI-Enhanced Future

Chapter 7
How to Have a Say in Shaping the Future of AI

Last month, I found myself watching a clip from a school board meeting in another state, where parents, teachers, and administrators had spent hours debating whether to "allow" AI in their classrooms. What struck me most wasn't the disagreement about AI, but how everyone seemed to be having entirely different conversations about the same topic. It illustrated a theme I've shared throughout this book: the term "AI" is so broad and all-encompassing that it can create parallel discussions, even among well-informed groups.

If you've ever sat through a similar meeting or asked school leaders about their AI policy (if they have one at all), you know that sinking feeling of everyone talking past each other. This scene is playing out in educational settings across the country as schools scramble to develop AI-specific policies. But without clear guidance, a solid understanding of what AI actually is, or the necessary training for educators, it's an uphill battle. This is especially true when you consider that our kids are already far ahead of us in their AI fluency, and Big Tech representatives have spent years cultivating relationships with education decision-makers.

It's like watching people debate whether to allow "vehicles" in the school pickup line without distinguishing between bicycles, cars, or eighteen-wheelers. There are genuine reasons to be concerned about AI technology—we've covered several of these issues in Part Two. But we're also looking at a future full of opportunity that we can shape.

What's clear to most families I speak with is how we've gotten stuck debating abstract and distant concerns about AI rather than tackling the practical realities of right now. While we should consider future issues, the future is shaped by the work we do today. So in this first chapter of the final section of the book, I'm going to suggest ways we can more concretely dig in and have a say.

We've Seen the Future Before

We sometimes forget that disruptive periods of innovation (especially those that might impact kids) are not new. In 1998, President Clinton announced funding to accelerate the introduction of the Internet into schools, saying that his administration would ensure that every child in America could access "computers and the information superhighway."[1] The communication struck me for how much it sounded like President Trump's 2025 executive order (EO) calling for AI literacy in schools.[2] Not only were both EO's about technology and bringing it into the classroom with urgency, but the response in many ways has been the same with equal parts enthusiasm and skepticism.

Yale Professor David Gelernter was a notable voice of dissent at the time of President Clinton's announcement.[3] He openly warned in an op-ed for *Time Magazine* that "skill-free children are overwhelmed by information even without the Internet" and called the Web "a propaganda machine for short attention spans." Reading Gelernter's warnings today is fascinating because he's not wrong. He rightly identified one of the most negative consequences of opening the digital floodgates. So we would be smart to learn from the past and use it as a map to guide us toward the future.

The difference between the Internet and new AI systems is that the Internet's arrival into our lives was very gradual. The speed of AI development today was hastened by the Internet, which opened the door all those years ago to the platforms that would help create today's AI. We're talking about a current reality built upon that of the past.

It Starts With Good Policy

All of this means that we need smart, supportive, realistic policies that support kids right now, lay the foundation for a strong future, and consider past best practices in addressing change. Families play a critical part in steering the ship. I speak from experience too—as a parent and policy advocate. Through my work supporting the New York State Education Department and New York City Schools on data privacy and AI policy issues, I've seen how much administrators value family input. They take stakeholder feedback, concerns, and ideas seriously and want more voices in the mix.

The current policy landscape feels chaotic because it is. Some schools have embraced AI as a transformative tool, while others have banned it entirely. Many districts have vague policies that leave teachers and students confused about appropriate use and allow unfair variability between schools. This patchwork approach creates real challenges for families trying to support their children and stress for kids navigating the rules. This is why your voice matters.

What You Will Discover in This Chapter

Throughout this book, I hope you've gained confidence in your own AI literacy and are ready to put your thoughts, beliefs, and ideas into action. Unlike previous chapters, which focused on understanding the issues and managing AI discussions within your family, this chapter will show you how to have a larger influence on your community.

You'll find tools to help you engage with schools, districts, and policymakers as an informed advocate for balanced, evidence-based AI policies. We'll cover how to decode school AI policy language, better understand what policies actually accomplish versus what they claim, and explore how to support educators navigating AI's ascent into schools. You'll also discover how to weigh in on academic integrity guidelines and push for more voices in policy development. And we'll look at how to get more involved at the local, state, and federal levels.

The goal isn't to become a professional advocate but to develop the skills and confidence required to engage meaningfully with the AI policy decisions that impact your child's education. Remember that you are equipped for this role simply by being an adult responsible for a child. It's the only credential any of us needs to advocate for policies that support kids. The time to do this is now—and your child's school is the best place to start.

Weighing in on AI Policies in Schools

The confusion around AI policy in schools right now is frustrating parents and educators and stressing out our kids. But moments like these also offer a powerful opportunity for stakeholders to come together and figure things out. When you consider how many schools are still scrambling to decide their approach to AI, family input on how to make this happen is usually more than welcome.

Families bring additional context and useful perspective from watching our kids navigate AI technology at home. Whether it's using AI tools for homework or playing games and exploring new features for fun, we can contribute insights that help schools. As with any challenge this complex, there are no clear and easy answers, but a rich and diverse set of voices can make all the difference.

The Role of Digitally-Eager Adults

While we like to refer to the youngest generations as "digital natives,"we tend to omit the part where adults made them this way. Generation Z and Alpha may have enthusiastically embraced the digital realm that generations before them built, but they certainly weren't born with smartphones in hand. We instead handed them these device and off they went to outsmart us.

Children in K-12 schools and colleges have also been fed a steady diet of technology in the classroom for years. And when children were removed from the schools during the COVID-19 pandemic and forced to navigate remote learning, it opened the floodgates to hundreds of new apps and edtech platforms. According to Learn Platform by Instructure, an average of

2,982 edtech tools were accessed per school district during the 2024–25 school year.[4] This compared to just 851 in the school year before the pandemic.

Consider the number of Chromebooks and iPads in schools these days as well—it's hard to picture a classroom without them. These notebooks flooded US schools during the pandemic.[5] The most reliable estimate from 2022 puts the total number of Chromebooks in schools at about 50 million, and we have little reason to think that number has done anything but grow.

If you consider this growth alongside the number of articles expressing "outrage" over kids "cheating" via ChatGPT, it should follow that adults bear responsibility for any such issue. We created voracious consumers of technology, avoided providing clear policies and reasonable guidance, and didn't adjust educational approaches in light of generative AI fast enough. We brought kids to this point, and now it's our responsibility to smooth things out, not condemn kids for fully embracing new technology and finding clever ways to use it to their advantage.

How Poor Policy Creates a Blame Game

For all the schools happy to welcome in more edtech tools, it's outrageous to consider how ill-prepared many have been to think about the impact of generative AI—including how kids may make it work to their advantage. The result is enormous stress being placed on children right now.

Students worry that peers who use AI more extensively are gaining an unfair advantage, while families worry that if kids don't use AI, they'll be hurt in the future. Students are so worried about being called out for "cheating" with AI that they're purposely adding typos and other quirks to AI-assisted writing. Even worse, some intentionally insert bad grammar into original work with no AI support because they worry that a detector will incorrectly flag their work.

In May 2025, *The New York Times* shared the story of a college sophomore who received a zero on a writing assignment after being accused of using AI.[6] To fight the accusation, the student was forced to produce a 15-page time-stamped PDF showing her work in detail, plus a 90-minute YouTube video documenting her writing process. This is absurd, and yet it's happening across the country with regularity.

If you Google "AI detector lawsuits," you'll see where a lack of strong, clear policy has gotten us—straight to lawyers with new business. It's also meant the creation of the "AI detection" industry; once nonexistent, it's now expected to be worth more than $2 billion by 2030.[7] This is what happens when uncertainty reigns—commercial interests scramble to monetize confusion and fear.

But there's time to right the ship, and families can play a critical role in ensuring kids aren't taking shortcuts to their education while also ensuring they aren't being penalized by a world we as adults have created.

What Good AI Guidelines Look Like

Strong AI academic integrity guidelines share several characteristics that separate them from the rushed, reactionary policies we've seen from many schools.

First, they focus on learning outcomes rather than tool restrictions. Instead of blanket "no AI allowed" policies, they explain when AI use enhances learning and when it undermines skill development. They help students understand the difference between AI that builds understanding and AI that replaces thinking.

Effective guidelines provide specific examples rather than vague principles. Students shouldn't have to guess whether asking ChatGPT to explain a confusing concept is acceptable while using it to write their conclusion paragraph isn't. The more specific the examples, the better students can apply the principles to new situations.

These guidelines acknowledge that AI capabilities evolve rapidly and include mechanisms for regular review and adjustment rather than treating academic integrity as a fixed set of rules. They include transparent processes for handling new challenges and appeals when students disagree with academic integrity violations.

Most importantly, effective guidelines connect AI use back to broader educational goals. They help students understand that academic integrity isn't just about preventing cheating—it's about fostering the thinking and character that will serve them throughout their lives. They recognize that learning to work thoughtfully with AI is itself an essential skill for the future.

How Families Can Help With School Policy

There's no better time than right now to dig in. All you really need to do is start asking the right questions.

Start With Your School's Current Reality

The simplest place to start is often the most obvious: ask school administrators about their current AI policy and how they're thinking about issues such as academic integrity. This conversation alone will tell you a lot. Some schools have detailed frameworks already in place, while others are still figuring it out alongside everyone else.

Find out who's actually making these decisions—sometimes

it's at the district level, sometimes individual schools have flexibility—and ask if there are upcoming meetings or surveys where families can weigh in. Be bold. You have every right to ask, and if schools have it all under control, they'll be able to provide quick and clear responses.

TALK TO TEACHERS

Your child's teacher is worth talking to as well. While teachers are typically stuck enforcing policies they didn't create, many have thoughtful perspectives about what works in practice. Teachers see firsthand when AI helps a student grasp a difficult concept versus when it becomes a crutch that prevents real learning. They also likely have opinions about what AI systems might make their jobs easier and allow them to focus more intently on the kids.

CREATE OPPORTUNITIES FOR INPUT

If your school doesn't have a formal process for gathering input, suggest they create one. School boards tend to be more receptive to family voices than individual administrators, especially when policy is still being developed rather than defended.

When you do get the chance to provide feedback or ideas, bring concrete examples from your own home. Share specific stories about moments when AI genuinely helped your child understand something complex, and contrast those with times when it short-circuited the valuable struggle of working through a problem independently. Real family experiences carry more weight than abstract concerns.

PUSH FOR PRACTICAL IMPLEMENTATION

Push for practical clarity about implementation. How will teachers distinguish between helpful AI assistance and inappropriate shortcuts? What training are educators receiving to apply these policies consistently? Since AI capabilities evolve monthly rather than yearly, how often will the school review and update their guidelines? Who gets included in those ongoing discussions?

Ask About Existing AI Systems

Don't forget to ask about transparency regarding AI systems already embedded in your school's technology. Which educational platforms use AI features? How does student data flow between these systems? What options exist for families who prefer minimal AI interaction? These questions help you evaluate whether policies are genuinely useful or just box-checking exercises.

Include Your Children in the Conversation

This process also gives families a natural opportunity to talk to children about their thoughts and experience with AI in schools. Ask them what tools they're encountering in class and whether they can tell when they're interacting with AI systems. These conversations help kids develop the critical thinking skills necessary to work thoughtfully with AI throughout their lives.

The students sitting in classrooms today will be the leaders, educators, and policymakers of tomorrow. Teaching them to participate thoughtfully in these discussions now is just as valuable as getting the current policies right.

Advocating for Student Involvement

In Part Two, we discussed the mental health and social development benefits of giving kids autonomy and a say over AI development and use. Now we'll talk about why it's critical to involve children in creating the technology policies that most directly affect them.

Children are the ones actually living with these rules every day and navigating the uncertainty. But kids can also start developing the judgment that will serve them well throughout their lives and begin becoming the leaders we need them to be. Many ethical questions will emerge in the years ahead that will require strong, capable, and thoughtful critical thinkers. The complicated questions around how to live alongside machines won't go away. We need the youngest generation to be ready.

If we exclude children from policy discussions, we're not just missing an opportunity to empower them—we're ignoring our own best interests in developing them into leaders.

Finding the Win-Win Policy on Development

Children are living with AI uncertainty in ways that adults can only theorize about. We simply don't have the context that matches their current lived reality. Children are figuring out in real-time what kinds of AI assistance genuinely help their learning and dealing with the daily stress of unclear rules and competitive pressure. They're in the center of the storm and have plenty of thoughts about what it takes to get out.

Children have also been more relaxed about AI systems. To them, these are just technologies representing "more of the same" in a world that's always been digital. This perspective can serve as a valuable counterbalance to adults rushing to dystopian future scenarios. Researchers studying the way kids think about AI found that at even 9–14 year-olds, children were able to communicate a much deeper, multifaceted under-standing of AI technology than even many adults.[8] Kids really "get" it. By not capitalizing on this fact, we risk a widening divide between educators and students—and in society by generation.

When students are involved in policy development, they bring practical insights that adults often miss. They know which rules actually work in practice and which ones students can work around. They understand the social dynamics of AI use in their peer groups and recognize when AI use interferes with their learning. More importantly, when involved, kids are motivated to create guidelines that support genuine learning rather than just forced compliance.

There are places seriously and successfully considering kids impact right now. For example, Scotland has taken the lead in creating a children's rights approach to developing policy.[9] In 1996, the Children's Parliament was launched to support implementation of the United Nations Convention on the Rights of the Child (UNCRC). The effort has already played a significant part in shaping AI policy through large-scale surveys of Scottish youth, framework development, and materials for teachers. Scotland's work is proof of this "win-win" strategy. The effort is good for kids, great for schools, and part of a larger effort in the country to lead in AI. With the June 2025 announcement that the most powerful supercomputer in the UK will be housed at the University of Edinburgh, their strategy may just pay off.[10]

Building Student Leadership Skills

Including students in AI policy discussions also helps them develop key leadership and critical thinking skills. When we ask students to wrestle with questions about appropriate AI use, we're helping them develop the ethical reasoning capabilities that will serve them throughout their lives. We also reinforce the notion that humans, not machines, are in charge—and that we're passing the baton to our kids.

Encouraging student leadership on AI efforts can also result in better inclusion and enthusiasm for our kids to join these efforts. At London Metropolitan University, students and faculty collaborated on a comprehensive framework for integrating AI into coursework with a directive to pay particular attention to how to better engage diverse groups of learners.[11] The results were impressive, showing a 20-30% increase in course module completion, better attendance, and greater satisfaction among students and teachers.

Students who feel ownership become the best advocates and eager stewards of good policy. Rules feel less arbitrary when you've been part of the team creating them—and kids are then more likely to encourage peers to do the same. Student ownership creates a natural partnership that ultimately benefits everyone involved.

What Student Involvement Can Look Like

Advocating for student involvement in AI policy development shouldn't be difficult. A great start is to recommend student representation on committees or task forces developing AI policies at your child's school. These shouldn't be token positions either. Students need real input into policy decisions, not just opportunities to provide feedback after adults have made the important choices.

Families can push for student focus groups or forums where young people can share their experiences with AI tools and discuss what kinds of guidelines would actually support their learning. These conversations should happen before policies are drafted, not after they've been written.

We can encourage schools to pilot AI policies with student input rather than implementing final rules without testing them. Students can identify problems with proposed guidelines and suggest modifications based on their actual experiences using AI for academic work—getting the policy right the first time.

Some schools are creating student AI ethics committees that meet regularly to discuss emerging issues and provide ongoing input as AI capabilities evolve. This ongoing involvement recognizes that AI policy isn't a one-time decision but an evolving conversation covering philosophical and ethical challenges.

Student contributions shouldn't be age-dependent either—all kids can participate in ways appropriate to their grade level. Elementary students can engage in discussions about basic concepts like honesty, effort, and the difference between helping and doing work for someone. Middle schoolers can tackle more sophisticated questions about learning goals and authentic assessment. High school students should be full participants in AI policy discussions.

Empowering Kids to Get Involved

One of the best ways for families to influence AI policy is encouraging our own kids to raise their hands and actively participate. We need to be cheerleaders and supporters of our children's efforts to take leadership positions.

Start by working with your child to identify places where they already have influence and can add AI policy to the agenda. Student government initiatives and other school governance organizations where students already participate are natural starting points. Another avenue is integrating AI discussion into debate clubs and similar forums where kids can work through the complexities of AI development and ethics.

It's not necessary to create new opportunities when so many already exist for kids to exercise leadership and include their voices in policy development. Thinking critically about technology's role is relevant to nearly everything a child does during their school day and in activities afterward. Fitting technology discussions into existing frameworks also helps develop critical thinking skills, ethical reasoning, communication, and other capabilities needed to navigate complex questions about AI use in professional settings.

Most importantly, students who participate in AI policy development today are preparing to lead thoughtful conversations about AI's role in society tomorrow. Armed with new confidence and understanding of the roles families can play, each of us can take the next step and start to exert influence over larger, more sweeping policies at the local, state, and federal level.

How to Participate as Governments Regulate AI

When I tell parents they can influence AI policy beyond their child's school, I often see that familiar "like I have time for that!" look. The idea of engaging with state legislators or federal agencies feels like a step too far when we're already navigating homework policies, parent-teacher conferences, and our own learning about AI technology. It's also intimidating. Our first instinct is that policy advocacy is full of hostile, messy, gridlock-filled domains unwelcome to new participants.

But this may be where families have some of the more lasting impact on our kids' future. And it doesn't require becoming a professional lobbyist. As someone responsible for children, you have as much right—and arguably as much expertise—as anyone involved in crafting policy related to kids. Legislators regularly acknowledge that the most useful insights come not from industry or technical experts but from those directly affected by an issue. They want to hear from those who will be most impacted.

It's not an all-or-nothing proposition either. The most effective family advocacy starts small and builds outward as you gain confidence. Think of it like learning to drive—you start in a parking lot before moving onto the highway.

AI policy decisions are being made right now at every level of government, often without meaningful family input. When families participate in these conversations, policymakers are better able to "stress-test" proposals and identify unintended consequences before draft policies become law. Research on public participation in technology governance consistently shows that when diverse stakeholders—including families—are involved, policies are looked at as both more legitimate and more effective in practice.[12]

It's also worth remembering that most policymakers are not computer scientists. Many are wrestling with the same questions that families struggle to answer. What does AI mean for children's privacy? How will it affect learning and future opportunities? What safeguards are needed? By contributing your perspective, you help ensure that rules are built around lived reality, not just abstract principles. Without the right context, good policy isn't possible.

Local Government Impact Opportunity

After engaging at the school level, the next step is local government. Cities and counties are making decisions about technology infrastructure, privacy protections, and digital equity programs that directly impact how families experience AI. In fact, more than 20 leaders representing cities from across the US have already convened as part of the National League of Cities to advocate for AI policies that support the autonomy and influence of local stakeholders.[13] These types of committees actively seek input from citizens directly impacted by local policies.

Starting locally might be as simple as attending one public meeting, submitting written testimony, or volunteering for a citizen committee. You don't need to come with expertise, just an authentic voice representing your family's and community's concerns. Local governments guide local policies around AI integration and manage related issues that matter.

For instance, It's difficult to have a conversation about AI technologies if we don't recognize the need for robust broadband coverage and community technology initiatives that ensure access. Consider that in 2021, a Pew Research Center study found that nearly one in four parents reported that their child struggled to complete schoolwork due to unreliable internet access.[14] These ancillary issues also require our support and represent a larger universe of policy decisions that are ineffective without thoughtful input from families.

Getting Involved at the State Level

With local experience under your belt, state-level engagement becomes far less daunting. States play outsized roles in AI governance because they set education standards, fund technology initiatives, and increasingly pass privacy legislation.

State education departments are also increasingly writing guidance for how AI should be used in classrooms. I have seen firsthand how parent perspectives shape policy discussions around AI, data use, and student privacy at the state level. It's a great place for families to have a say.

Parents can also affect state policy by becoming more active with established advocacy organizations. Groups focused on student rights, privacy, or equitable technology access often provide training, talking points, and logistical support to help families participate effectively in state-level advocacy. By connecting with these organizations, you amplify your voice and benefit from collective expertise.

Leveling Up to Provide Federal Input

At the federal level, the issues may feel more distant, but the impact is far-reaching. Federal frameworks shape how AI is tested, how companies collect data, and what transparency requirements apply across industries. Families are frequently recognized as key stakeholders in federal policy development.

The US Department of Education's 2025 guidance on AI in schools emphasized the need for parent engagement in shaping both AI policy and its implementation.[15] Families can weigh in during federal comment periods, which are open to anyone, or work through national organizations that aggregate parent concerns.

Congress is also considering broader legislation to regulate AI. Hearings in 2024 and 2025 have included testimony from educators, students, and families, illustrating that policymakers are receptive to voices outside the tech sector. While it may not be realistic for every parent to testify in Washington, even submitting written testimony or joining advocacy campaigns coordinated by family-centered nonprofits ensures that parental perspectives are represented in the debate.

Federal funding decisions also shape the priorities of AI research and innovation. Billions of dollars are being directed toward AI research each year, and advocacy can help channel those funds toward educational uses, child safety, and accessibility. Families have a unique role in reminding policymakers that technological innovation should serve human development, not the other way around.

Becoming an Advocate, One Step at a Time

You don't need to tackle every level of government at once. Choose the arena that feels most manageable first. Perhaps sign up to receive a state education newsletter, attend a local task force meeting, or respond to a federal comment request. Each small action helps build confidence and understanding. And every interaction adds to the pool of parent voices that policymakers draw from when making decisions.

Your perspective as a parent navigating AI in daily life is not a distraction from the "serious" policy conversation—it is the grounding force that keeps policy tethered to reality. By sharing your experiences about how AI affects homework, privacy, or digital equity, you help ensure that the resulting policies are workable, fair, and protective of children.

AI technology will not slow down to give policymakers time to catch up. The governance structures being put in place today will define how children experience AI for decades to come. Families who engage in these conversations help ensure that the rules reflect human priorities of fairness, safety, opportunity, and trust.

The daily challenges of raising children in an AI age make you exactly the kind of expert these conversations need. In my experience, the best part of this type of participation is the chance to meet like-minded, smart, passionate families who are really making a difference. For me, seeing grassroots organizations at work has been an incredible learning experience and a connection I'd recommend every family consider exploring.

Getting Involved at the Grassroots Level

When we start to see meaningful policy shifts at the federal or state level, it's usually because groups of families and community members have been laying the groundwork for months or even years. "Grassroots" is a term covering this "bottom-up" approach where communities lead advocacy efforts, providing a counterbalance to lobbyists and other well-funded entities. It's typically those most affected joining forces to be heard, and it can be a very effective approach.

The power of grassroots advocacy lies in its numbers, shared objectives, and the kind of authentic storytelling and lived experience that cuts through political noise. It's about building a community base of knowledge and supporting momentum that makes sweeping changes possible.

Grassroots work on AI issues can happen everywhere: through PTAs, local nonprofits, community organizations, neighborhood groups, and even informal parent networks. It can also be an add-on to other issues and organizations with long-standing support and well-honed infrastructure, such as educational equity organizations, data privacy efforts, and children's rights advocacy groups.

Getting involved at the grassroots level gives families a chance to learn best practices around advocacy while benefiting from diverse skills, expanded networks, and influence that can deliver surprisingly powerful results. You're not starting from scratch or working alone. You're building on work that's already been done and learning from people who've been at policy advocacy longer. It's a great way to connect with like-minded neighbors who share your concerns and make new friends from around the country and world. You get access to resources, understand strategies that have worked, and benefit from the kind of context that achieves results.

Starting Local With Community Groups

Your child's school and local community are the most natural entry points for grassroots work. PTAs and parent councils are ideal for conversations about AI policy because they have the focus, resources, and established protocol for influencing school decisions. Parents can use existing PTA infrastructure to initiate discussions on responsible use, family expectations around academic integrity, or how AI impacts digital citizenship.

It can be as easy as showing up to PTA meetings or going farther and proposing a session at a meeting, working group, or parent night. PTAs welcome this participation. As a former PTA president myself, I can tell you that these organizations truly value new voices and fresh thinking to address issues and challenges that affect a school's community.

Informal conversations matter too. Small parent discussion groups—over coffee or online—often lead to insights and collective approaches that benefit entire school communities. When families feel supported and informed, they're better equipped to advocate effectively, whether that means asking questions at board meetings or requesting clearer communication from administrators. They're also empowered to talk among one another and develop ideas, questions, and solutions.

Local libraries are also becoming increasingly important as hubs for digital literacy support. Libraries serve as critical meeting points for communities and often play a key role in connecting and educating neighbors of all ages. A report from the University of Albany found that increasingly libraries are offering essential AI workshops and related forums, making them trusted, neutral spaces to start working on solutions to issues families are worried about.[16]

Any place your community comes together is somewhere to

look for groups that may be addressing AI innovation. You're likely to find like-minded families who are also hoping to team up and advocate for kids.

Benefits of Larger Advocacy Efforts

Beyond your school community, hundreds of child- and education-focused nonprofits are already wrestling with AI literacy, policy development, and how AI should show up in classrooms. Some focus specifically on digital issues. Others center on education but make AI and edtech core to their work. Either way, these groups offer families a way to connect individual concerns to organized advocacy that's already happening.

These include familiar names such as The Family Online Safety Institute and Common Sense Media, which provide leadership and advocacy on issues related to kids and technology. On the general schools and education front, there are organizations such as the National Parents Union, which provides materials, research, and other support.[17] The group has included technology issues in their 2025 agenda, including enhancing digital literacy, citizenship, and safety; bridging the digital divide; supporting solid policies around AI in education; and promoting social media literacy.

I've also worked closely with the Parent Coalition for Student Privacy, which focuses on data privacy in schools and compliance with data privacy laws. They provide news and insights via their blog and podcast regarding how edtech and now AI tools can materially affect our children's legal right to privacy.[18]

How Established Efforts Generate Momentum

Data privacy, digital rights groups, and organizations tackling "ethical AI" are great entry points to consider in your grassroots exploration. While these groups may have more experts, such as technologists or academics, among their ranks, they welcome all who are interested in addressing these issues. There's enormous breadth in offerings, and sometimes even just signing up for their newsletter or downloading materials can be beneficial.

There are grassroots efforts dealing with issues of transparency, algorithmic bias, and fairness, such as the Algorithmic Justice League. All Tech Is Human convenes academics, educators, and technologists to co-create resources for responsible technology use and holds events in many cities. The Mozilla Foundation has run events such as the Responsible Computing Challenge and MozFest, which bring communities together to address digital equity and AI literacy.[19]

The list is long, but what all of these organizations share are resources, tips, materials, ways to get involved, and a network of other advocates. They also often offer workshops, webinars, and resources designed for non-experts. These sessions typically generate practical strategies, in-depth insights, and relevant facts and figures that families take back to their local communities.

Making Your Voice Count in Every Way

Whatever format you choose, the key is bringing authentic family stories into larger conversations. Policymakers, technologists, and educators often lack the daily perspective of families watching children navigate AI tools at home. Your specific examples—the confusion, the breakthroughs, the boundary-setting struggles—anchor abstract discussions in reality.

Document what's happening in your family. When your child uses AI to help with homework, note what worked and what didn't. When school policies create confusion or seem out-of-touch with technology being used, write it down. These stories become powerful tools for advocacy.

Building capacity over time matters more than any single action. The networks, knowledge, and relationships you develop today will help your community navigate whatever AI developments come next. Research on civic tech movements indicates that the most effective grassroots work happens through sustained participation that builds trust and infrastructure before time-critical issues arise.

Grassroots engagement also teaches your children what civic participation looks like in the digital age. They see that technology isn't just handed down from corporate or government decision-makers but something communities can shape together. They learn that engaged citizenship now includes asking hard questions about AI and insisting that innovation serves human values.

There are also grassroots organizations that include kids in their ranks and see the value of teaching children to be the kind of advocates we need in the future.

The grassroots level is where abstract policy becomes lived reality. By stepping into these conversations locally, you help ensure that reality reflects your community's values and serves children's best interests. You're not just reacting to changes— you're helping create them.

Key Chapter Takeaways

The most important thing I hope you take from this chapter is that you don't need to be a technology expert or policy professional to have a say in how your children experience AI. You just need to ask questions, push for approaches that actually help kids learn, and work with educators and policymakers. All families are uniquely positioned to do this by the nature of being responsible for a child.

Families are crucial when it comes to policy affecting kids, though sometimes we need to nudge our way into having a seat at the table. Understanding how much your voice matters to this process should give you a sense of control over what happens next. And hopefully it also means that you encourage your children to come along and find their voice as they prepare to lead us into the future. It will take all of us, and there's no better time to start than now.

Stepping in and Helping Schools

AI policy development is messy right now. Rules are inconsistent by state and even by school within a district, changing rapidly, with nobody having clear answers. But this chaos also creates an opportunity for families to have influence. Unlike education policies that have been locked in place for decades, AI policies are still being written based on what communities tell policymakers and how things actually work in practice.

Educators want families involved. Last year the National Education Association, the country's largest union, approved a policy statement calling for policymakers to "actively include the voices of diverse students, educators, and caregivers when adopting AI-enabled technology and creating AI-related policies."[20] Many government and advocacy organizations are suggesting the same.

Throughout this book, we've looked at how to decode school AI policies—from blanket bans that ignore how AI could actually help kids learn to vague rules that leave everyone confused. We've also explored related issues like data privacy that make AI decisions even more complicated. Understanding these patterns provides a roadmap for families to follow.

When policies use meaningless terms like "appropriate use" without explaining what that means, you can demand specific examples and real explanations that teachers and students can actually follow. Your daily experiences with AI—how it affects homework time, raises questions about fairness, or starts family conversations about technology—are precisely the kind of real-world information policymakers need. These lived experiences keep abstract policy discussions grounded in reality. Without them, policies won't actually protect our kids.

Teachers, families, and policymakers all want kids to succeed. Building strong partnerships takes patience and a willingness to understand different perspectives. Teamwork can make the path to fair and effective AI adoption a success.

From the Classroom to Capitol Hill

We've explored how AI policy advocacy works at every level, from individual classrooms to federal legislation. Each level carries different opportunities and challenges. Local engagement is usually the most accessible starting point—your voice carries real weight, and relationships are easier to build. State-level advocacy can create bigger impact while still allowing meaningful family participation. And while federal engagement might seem distant, it shapes the larger frameworks around everything else.

The most effective advocates understand how these levels connect. Supporting federal legislation that enables strong state privacy laws. Pushing for state funding that supports local digital equity programs. Making sure local policies align with helpful state and federal standards. This creates more comprehensive impact than focusing on just one level.

Effectiveness of Grassroots Work

Grassroots involvement is where abstract policy becomes real life. Whether through PTAs, local nonprofits, community organizations, or informal parent networks, grassroots work lets you collaborate with neighbors who share your concerns, educators who teach your children, and local leaders who understand your community.

The connections and relationships you build through grassroots work create lasting change that goes beyond individual policy battles. These networks become the foundation for tackling future challenges while showing your kids what civic participation looks like in the digital age.

Your Family as Policy Leaders

AI policy advocacy isn't a one-and-done activity. Technology keeps evolving rapidly, policies need regular updates based on how they actually work, and new challenges keep emerging that need family input. The skills and relationships you develop through this work will serve your family well as AI continues changing education and society.

This doesn't mean becoming a full-time activist. It means staying informed about developments that affect your kids, maintaining relationships with educators and other families who share your concerns, and being ready to engage meaningfully when opportunities arise. Sustainable advocacy means picking your battles thoughtfully and focusing on areas where you have genuine interest and relevant experience.

We've also seen how students themselves need to be central voices in these discussions. When we include kids in AI policy development, we help them build the leadership and critical thinking skills they'll need as tomorrow's decision-makers. Advocating for student involvement isn't just about being inclusive—it's investing in better policies and preparing the next generation to make thoughtful decisions about AI.

You bring the only expertise required: responsibility for a child who needs your guidance. The future of AI in education doesn't depend on distant experts or corporate interests. It depends on families like yours who are willing to step up, speak out, and work together to get this right for all children.

Chapter 8
What Excellent AI Support at Home Can Look Like

The real work of preparing kids for the future happens at home. While families know this instinctively, the introduction of AI into our home lives has made it harder to feel confident about what we should do next. What makes the situation particularly frustrating is that, as adults, we have little to draw upon from our own experiences growing up to assure kids about what to expect and how to get there.

"I used to worry about whether my kids had enough pencils and notebooks," my friend Caleb told me recently. "Now I'm trying to figure out if I should help them set up ChatGPT accounts, and if I do, what privacy settings to use, and how to teach them when AI support crosses the line into cheating. It feels like each year I'm preparing them for a school year that bears no resemblance to even the one just before, and that there's no handbook that can be written fast enough."

These aren't just practical issues but also deeply personal ones that touch on how we want our children to develop as learners and grow as people. The decisions we make about AI at home will shape not only our kids' academic success but also their relationship with technology, confidence in their own abilities, and understanding of what they can do with it all.

But if we face the challenges of this uncertain period head-on, it can also be an opportunity to clear the decks, rethink the rules at home, and build better habits. The uncertainty many families feel about AI isn't a sign that we're unprepared—it's a sign that we're taking this transition seriously, and that's a good thing.

Building on What You've Learned

Throughout this book, you've gained confidence to understand AI from multiple angles. You've learned to think critically about AI systems, understand privacy implications, and engage with policy discussions at school. You've wrestled with questions about bias, explored the educational potential of AI tutoring, and considered how these tools might reshape not just homework but how we think about learning itself. Now it's time to get practical about supporting kids at home and what you should consider in crafting your family priorities.

The difference between understanding AI and successfully integrating it into a family's routine is like the difference between reading about riding a bike and actually getting on one. You need specific strategies, realistic expectations, and the willingness to adjust as you go. And it's never fully about the bike either. A bike is simply a way to exercise, spend time outside, or be together as a family. Like AI, it's a means to an end but not the end in itself. The technology isn't the story here —we are.

If a family values creativity highly, then the question becomes how to use AI tools to support rather than replace your child's original thinking. If independent problem-solving skills are priorities, then it's about how AI assistance can strengthen rather than weaken those skills. These questions don't have universal answers, but they have answers that are right for you. AI is a tool—the question is how and why to use it at all.

What You Will Discover in This Chapter

In this chapter, we'll transform the AI knowledge you've gained into school-ready and at-home strategies that work for your family. We'll start with setting up accounts, encouraging good habits, and deciding on individual evaluation criteria for choosing AI tools and systems.

We'll look at what it means to foster belief flexibility, tolerance for change, and independent thinking with your kids. In a world where AI capabilities evolve rapidly, children who can adapt their thinking and stay curious will thrive. This doesn't mean accepting every new technology uncritically, but rather developing the mental flexibility to evaluate changes thoughtfully and adjust accordingly. And this includes having the wisdom and humility to admit when you may have been wrong or have changed your mind.

From there, we'll explore how to encourage the development of "Emotional Intelligence," or "EQ" as it's come to be known. We'll also look at why interdisciplinary study should be considered as a path for kids to take at college and university. This isn't about adding more to your child's plate but about helping them develop the emotional strength and mature thinking that will matter most. As AI handles more routine cognitive tasks, uniquely human skills like empathy, ethical reasoning, and creativity will become more valuable, not less.

Finally, we'll focus on helping children find their personal voice. This is the antidote to concerns about AI assistant writing. By cultivating and recognizing one's personal voice, all of us bring our personal perspectives and uniquely human, not machine-created, style to any work.

The most crucial goal here is to raise kids who can use these powerful tools without losing themselves in the process. The goal isn't perfection but progress. Every family's approach will look different based on personal values, your children's needs, and individual circumstances. What matters is developing intentional approaches that support authentic learning while preparing kids for whatever comes next. And that's something we are all already well-equipped to do as the guardians of developing little humans.

Setting up Accounts and Supporting Good Habits

When it comes to choosing educational support AI, the options are endless, with companies offering bold claims of successful outcomes. At the same time, there are few sources to guide families in making the right choices about what platforms to use and how.

The key isn't finding the "perfect" AI tool but understanding what actually works for your child's age, learning style, and your family's needs—and then trying it out. As the educational AI landscape expands from simple homework help to sophisticated tutoring support and everything in between, having clear family-specific criteria helps you quickly and effectively decide.

Evaluating for Quality and Avoiding Novelty

Before committing to any AI educational platform, it's important to treat the process like you would any other decision related to your kids and school—step by step. Start with free trials or basic versions to test functionality and age-appropriateness, especially for any AI systems that you are using at home. During trials, pay attention to how much time your child spends on the platform, whether they remain engaged with non-AI learning activities, and how the tool impacts their confidence and motivation. Consider whether the AI tools will grow with your child or become obsolete as they develop.

Think about how any tool or platform integrates with your family's existing learning routines. AI should enhance your approach to education, not dominate or replace it. The landscape is changing so rapidly that we need to make these decisions individually based on fundamental requirements that meet a family's specific needs. And if a platform isn't meeting those objectives, cancel your subscription— and also be sure to request that the company delete your data when you close an account.

Consider Your Child's Developmental Stage

Back to the bike metaphor—think of AI tool selection like choosing the right bicycle for your child. You wouldn't give a seven-year-old a mountain bike with complex gear systems, just as you wouldn't give a teenager training wheels. Each stage requires different features that should match where your child is developmentally.

For elementary school-aged kids, you want technology that feels more like a helpful learning game rather than a sophisticated conversational partner. Your seven-year-old shouldn't need ten minutes to figure out how to ask a question. Look for simple interfaces that encourage thinking through problems rather than providing answers.

Preteens, on the other hand, can handle more sophisticated

interactions while still requiring guardrails. At this stage, we're teaching kids to both use AI and think about what the technology means. Your twelve-year-old should be able to explain what they learned from an AI interaction too. The best platforms encourage curiosity while providing structure and support.

By high school, the focus shifts to teaching kids how to use AI tools responsibly. Teenagers should consider platforms that offer transparency regarding their capabilities and limitations, tools for citing AI assistance appropriately, and support in developing reasoning skills.

Privacy-First Approach to Managing Accounts

Remember when setting up your child's first email account felt like a big deal? You probably spent time thinking about privacy settings and wondered if you were giving them digital freedom too soon. Setting up AI accounts brings those same concerns and more.

The privacy considerations are significant. Unlike traditional learning software that just stores quiz scores, these tools retain conversation histories and learn from interactions. So an awareness of this new reality must be matched to a solid strategy in platform selection and setup.

Try starting with the most restrictive privacy settings available, then gradually adjust based on your comfort level and individual educational needs. It's much easier to relax restrictions later than to find out later what data has already been collected. For younger children, adults should manage the account directly, which is also legally required as children under 13 can't consent to platforms themselves. But teenagers can have more flexibility and control.

Most AI platforms include social features that allow for the sharing of conversations or creative work. Unless these are specifically needed for educational collaboration, it's probably a good idea to turn these features off initially (or just avoid using them if you can't disable the feature).

And finally, be sure to read privacy policies and terms and conditions to ensure you understand where any data is stored, whether it's shared with other companies, and what happens when you close an account. Also be sure to actively close accounts when you no longer need them—following up with an email to the company to delete your personal data.

Working New Tools into Existing Routines

Here's something parents tell me all the time: they love the idea of AI making homework less stressful, but they worry about losing control of the process. That tension makes perfect sense. Yes, AI can be smart, helpful, and efficient as a tutoring tool. But the process of learning something hard doesn't disappear just because there's new technology in the mix.

Before making any AI-related changes in your home, take an honest look at what's already working (or not) with your child's learning process. Think about homework routines, study habits, what helps your kid focus, and what sends them into frustration spirals. The fundamental principles of effective studying shouldn't change. AI should slip into existing routines as assistance, not replace what's already working or eliminate the hard work that real learning requires.

When you find an approach that clicks for your family, many AI platforms let you create "projects" with preset instructions so you don't have to reinvent the wheel every time. I've put together some examples and a template on my website, aiforfamilies.com, to help you get started if this feels useful for your household.

How to Keep Learning at the Core With AI Use

Now let's talk about the elephant in the room: will AI make learning too easy? Actually, I think the opposite is often true. Using AI well means your child has to ask better questions, think harder about the feedback they receive, and keep trying until something clicks. AI isn't taking tests for kids or debating on their behalf. The idea that these tools will somehow dumb down education misses how rigorous the process can and should remain.

Sure, a child can use AI to taken shortcuts to doing their homework. But what happens when their teacher asks a question in class the next day? Can they actually explain what they learned?

This is why encouraging kids to slow down matters so much. Make sure your child can explain AI-generated content in their own words. If they can't talk you through what they learned, they probably need more time with the material—which has always been true, long before AI existed.

Watch how your child's confidence, understanding, and academic performance shift as they experiment with different approaches. Be ready to adjust how much time they spend with AI tools, switch which platform your family uses, or modify routines based on what's actually working. Trust your instincts throughout this process. You know your child better than any algorithm ever will.

When an AI educational tool starts feeling like helpful assistance that supports your family's learning goals rather than a shortcut, you'll know you've found the right balance. As kids and families get the hang of working with these tools, staying flexible and tolerant of change becomes the ongoing work. A lot of the stress in the months and years ahead will come from continued disruption—that's inevitable. But by keeping learning central and your expectations consistent, you'll be able to navigate whatever comes next with confidence.

Building Emotional Intelligence for the Future

Many readers, and GenXers in particular, may remember Daniel Goleman's groundbreaking 1995 book *Emotional Intelligence: Why It Can Matter More Than IQ*.[1] "EQ" is commonly understood as the ability to recognize, understand, and manage emotions in oneself and others. Goleman argued in his book that these capabilities frequently outweigh raw intellect in predicting future career success.

At the time, the idea that one's EQ could matter as much, or even more than, an individual's IQ was provocative. But fast-forward to today, and researchers are finding that EQ is linked to job satisfaction, work performance, effective leadership, and overall well-being.[2] Goleman was more than right, he foresaw a skillset that would be critical in an AI-led world.

The ability to offload more technical or rote activity to AI systems has elevated the need to consider the new ethical and philosophical implications that come with it. EQ isn't just a "nice-to-have" anymore. Understanding nuance, reading the room, and motivating groups will serve as an important counterbalance to machines. Even those in more technical fields have started to recognize the need to cultivate these skills.

One of the most notable is former Google engineer Chade-Meng Tan, who pivoted to pursuing mindfulness in 2007. Inspired to consider the mind more fully, Tan gathered experts to create an internal course for Google employees called Search Inside Yourself (SIY).[3] The course became so popular that he later created started a nonprofit educational institute with the same name with the mission supporting practical mindfulness and emotional intelligence.

Children who combine emotional intelligence with ethical reasoning won't just thrive in future AI-driven workplaces; they'll guide AI systems toward outcomes that reflect human values. They will be sought after as leaders and take on important responsibilities as AI's influence grows. The added benefit of EQ is that it can make us all better people too—kinder and more empathetic friends, thoughtful members of a family, and more highly engaged citizens. It's a win-win for everyone and for the future.

How EQ Leads in an AI-Driven World

While cultivating EQ skills is valuable in any scenario, its impact on AI development and adoption—and our kids' job prospects—isn't some abstract concept. As AI systems automate technical tasks, human capabilities for motivating, collaborating, and resolving conflicts become essential. In fact, many of the top skills outlined as critical to the future in the World Economic Forum's 2025 *Future of Jobs Report* reflected high EQ capabilities.[4] Skills making the top 10 list include resilience, flexibility and agility, creative thinking, motivation and self-awareness, empathy and active listening, and curiosity.

Cultivating these skills goes beyond future career prospects as well, making them critical to helping shape how society addresses the technology's disruption. AI innovation has brought with it uncertainty and moral quandaries that need to be addressed in both its output and development. As AI powers and amplifies systems, like social media platforms, that already take an emotional and psychological toll on many users, we need people who can consider and address the consequences. It's about the entirety of how we live right now and how important it is to be emotionally, psychologically, and socially prepared.

A few ways that EQ can help right now include:

Avoiding AI-Amplified Echo Chamber

As we covered in earlier chapters, the challenge of succumbing to dangerous "echo chambers" is equally shared by adults and children right now and has grown as an issue thanks to social media. From amplifying tribalism, distorting news, and bolstering efforts by bad actors to manipulate young minds, there is a lot to tackle even now.

But by cultivating EQ skills, we can more specifically address this issue and help free ourselves and our kids from these cognitive and emotional traps. Simple activities like spending more face-to-face time with others, and especially with those we might not agree with, start to poke holes in these digital bubbles. This could mean encouraging kids to join debate clubs, playing outside more in person, participating in sports, and interacting with children from different schools or backgrounds.

For adults, it means changing up our news sources occasionally, spending less time on social media, and engaging with family, neighbors, colleagues, and even strangers who hold different views. These can be critical exercises that help us see people as complex humans, not simply opposing one-dimensional data points.

Opportunity to Cultivate Belief Flexibility

If you've ever watched your teenager passionately defend a position they held for a few weeks before discovering new information that entirely changed their mind, you've actually witnessed the beauty of "belief flexibility." What you may not have considered is how important this willingness to shift one's thinking will be to surviving and thriving in the future.

In an era where AI can instantly generate convincing arguments, where personalized algorithms create these powerful echo chambers we just discussed, and where information landscapes shift constantly, the ability to update one's beliefs when presented with better evidence is essential.

You might be thinking, "Wait, don't we want our kids to have strong convictions and principles?" I'd suggest we want them to have strong values, a strong moral compass, and convictions that are rooted in established facts. Because it's precisely our core values and personal beliefs that should make us flexible to new facts, alternate ideas, or different approaches to addressing complex problems.

Fostering a growth mindset about belief formation means updating beliefs when encountering better evidence. This is why AI systems should be thought of as a means for getting better information to formulate our beliefs, not letting the machine do the work of formulating opinions for us.

Families can play a key role in reshaping how we discuss current affairs, the importance of curiosity, and why we should build flexible personal belief systems. Consider using language that celebrates intellectual growth and a willingness to say you've changed your mind or feel you were previously wrong. These examples help children see that belief evolution isn't just normal but critical to human progress (and growing up).

Avoiding the Risk of Cognitive Offloading

When we let an AI system develop opinions for us, we aren't just risking getting stuck in echo chambers or being robbed of learning. We actively damage EQ. To offload actual decision-making means abdicating to a machine all of those skills we've talked about. We lose practice in reading nuance, weighing competing values, and sitting with uncertainty. We miss opportunities to develop empathy by considering how our choices affect others.

Most critically, we skip the emotional labor of wrestling

with difficult questions—the very process that builds our capacity for complex thinking and ethical reasoning. Researchers call this "cognitive offloading."[5] This practice impacts decision-making accountability and can make it easier to settle into a set of beliefs rather than take on the emotionally challenging work of arguing or defending our thinking with facts and detail. It amounts to intellectual laziness but can ultimately also result in the deterioration of these precious EQ skills.

Navigating Uncertainty into the Future

Perhaps the biggest big-picture challenge of AI and other new technology is the uncertainty it heaps upon us. From big questions to small ones, we need to get comfortable with not having the answer—and we must find a way to be at peace with this fact.

For instance, will AI lead us into danger or utopia? No idea. Will certain professions disappear? Who knows. But what we do know is that we have choice, personal agency, and a type of intelligence that cannot be replicated. Just knowing these things should provide us with the confidence to be okay with the resulting uncertainty.

In an AI-integrated world, the "right" approach to any situation may frequently be unclear and also change with time. So rather than rushing to provide definitive answers when children encounter ambiguous situations, guide them through frameworks that account for uncertainty. And then make sure you tell them it's okay.

The truth is, there's a lot in life we're uncertain about, but we also have the power to make choices that can keep us moving in the right direction—and that is core to EQ.

Modeling EQ as Adults

The most powerful way to teach belief flexibility, tolerance, and independent thinking is by demonstrating these qualities in our own behavior. Consider sharing with kids experiences where you changed your mind or a moment when you realized you might have been wrong about something. Admit to them when you don't know enough about a topic to definitively answer their questions, and show curiosity about what they think.

When you encounter new information that challenges your assumptions, involve your children in the thinking process and use it to engage in meaningful discussion. Show your kids how you evaluate new evidence, consider different perspectives, and sometimes update your understanding as you better grasp a situation.

In an AI-integrated future, the most successful people will be those who can adapt quickly, think flexibly, and remain open to new possibilities while maintaining their core values and ethical principles. They "take a beat" before responding, don't fall prey to tribalism, and are likely to even avoid social media and other divisive realms.

These EQ skills will serve your children well regardless of how technology continues to evolve or what they choose as their profession in the future. And most importantly, it will give them the confidence to thrive as independent thinkers who know who they are and can be the leaders we need in the years ahead.

Helping Children Find Their Personal Voice

To better develop EQ requires deeper self-awareness, and fundamental to this awareness is a solid understanding of what constitutes one's "personal voice." In many ways, a person's ability to responsibly work with generative AI tools is highly influenced by how confident a grasp they have on this personal awareness. It's the fundamental question of "who are you?" and the answer can be curative to any concerns we may have about AI and academic integrity or the technology's impact on creative fields.

AI can become a crutch when its overly bland and homogenized output is never fully considered in the context of how a person knows themselves and their individual style. This ultimately gets us to the core of the risks and opportunities inherent to using AI tools. If a child is never given the opportunity nor encouragement to find their personal voice, then AI will only handicap their learning, and this is what we should worry most about.

What is Personal Voice?

Our personal voice is how we "sound" or communicate. It's our "signature style," with all of its nuance, including our unique way of talking, telltale signs of imagination, and even the favorite words we use, time and again.

It's a uniqueness that I would argue has been overlooked as we've pushed toward conformity in writing for a long time, and well before generative AI burst onto the scene. We've focused so much on rules, structure, and "good" grammar that it has relegated anything unique to side work, like "creative writing" classes or clubs. We spend so much time competing and comparing ourselves to others that we forget to be the best version of ourselves and understand what that even means.

Who your child is and what makes their communication unique is now critical to getting kids started working alongside AI systems. And I'm not talking about the current trend toward inserting errors and strange choices into one's written work to ensure it doesn't "appear to be AI." This is about those natural identifiers and hallmarks of our child's individual personality. It's like when you can identify your child's artwork or essay without even seeing their name on the paper—it's this essence that we must harness and help kids grow.

No AI system can replicate what makes us unique, regardless of what anyone says. Any discussion of a system being able to "replicate the style" of a writer doesn't consider the author's ability to change styles and defy statistical expectations as they see fit. The stronger each of our personal voices is, the more sure we can all be that AI cannot take anything from us.

Our "Voice" is in Everything

Personal voice isn't just about writing style either; it's a holistic picture of all the ways a person communicates. It's about how your child sees the world, processes information, and expresses their unique thoughts, perspectives, and beliefs. It shows up in the questions they ask, the connections they make between seemingly unrelated ideas, and the way they explain complex concepts to others.

When your eight-year-old describes their weekend using elaborate metaphors involving superheroes, or when your teenager writes a history essay that somehow connects ancient Rome to modern social media patterns, they're developing voice. It's that distinctive way of thinking and communicating that makes our kids' contributions valuable in any collaborative setting—whether they're working with AI tools or human colleagues. And whether the execution is written or verbal, art or film.

It also doesn't have to be consistent either. Voice is not an identity; it's a highly personal and fluid style. Voice develops through authentic expression rather than imitation (although imitation is part of the journey of discovery). While children need to learn proper grammar, citation formats, and essay structures, they also need opportunities to experiment with their ways of communicating ideas through writing, art, performance, debate, discussion, and play.

Families have an important role in encouraging this journey of self-discovery, from noting when a child seems to be expressing themselves uniquely to what we think our own personal voice sounds and looks like as well.

Creating Space for Authentic Expression

How do you encourage the cultivation of personal voice? To begin this work, help children to identify what they're naturally drawn to and provide opportunities for them to explore those interests more deeply. A child fascinated by architecture will develop a different voice when writing about urban planning than one who's passionate about marine biology. Pay attention to the topics that make your children light up in conversation, the questions they return to repeatedly, or the activities they lose track of time doing.

Consider going analog too. That is to say, encourage putting pen to paper. The way your brain processes the written word is different from that produced on a computer. Research shows that writing by hand fires different parts of our brain, and that can be useful to our intellectual development.[6] Activities like writing by hand aren't just a great way to encourage the discovery of personal voice; they are mental conditioning activities too.

Create family environments where unusual ideas are welcomed and explored rather than immediately corrected or dismissed. When your child makes an unexpected connection or asks a question that seems off-topic, follow that thread and see where it leads. These moments of divergent thinking are where personal voice often reveals itself.

Finally, encourage your children to start creative projects with their ideas before involving AI assistance. When they're writing a story, help them identify what they personally find interesting about the topic. What experiences from their life connect to this theme? This personal investment becomes the foundation that AI tools can then help them develop and refine.

Using AI to Amplify, Not Replace, Voice

All of this is why it's so important to approach using AI with an empowered mindset rooted in self-awareness. When children use AI tools for creative work, we need to teach them to approach these tools as amplifiers of a personal idea rather than as a substitute for their original thinking. AI can help children explore different ways to express the same idea, suggest variations on their themes, or provide feedback on their creative choices. It shouldn't ever do the work for them.

Show children how to use AI to push their thinking further rather than provide a lazy way out. If a child is writing a story, they might ask AI to recommend different endings for them to evaluate and pursue or to help them think through the motivations for characters they've already established in their mind. The key is maintaining ownership of the creative vision and process while using AI to explore and refine that work.

Helping children recognize the difference between AI-generated content that reflects their voice and content that sounds generic or disconnected is crucial here. We've spent too much time talking about "AI detection" and "sounding like AI," but not enough about what it means to sound like oneself.

Celebrating Unique and Individual Perspectives

Even at a time when we appear to celebrate individuality, it can take the shape of "identity," rather than the "one-of-a-kind" type of uniqueness I'm advocating for here. Helping children see what makes them unique as individuals and not as part of any one "group" also has implications for the work we've discussed in breaking free from echo chambers and other AI-amplified controls.

In an AI world where creating algorithmic consensus is increasingly easy, children who can offer authentic, well-reasoned arguments and alternative perspectives will be better prepared to stand out and fight for better. But it's a muscle that needs to be flexed. Families should seek opportunities to work out together what it means to be an individual.

Kids need to also become the champions of their own unique communication style and learn to advocate for what makes their contribution different. In a world where AI can generate endless variations of "acceptable" responses, children who can confidently articulate why their perspective matters will be the ones who shape the conversation rather than simply participate in it. They can serve as the rebels that save us all from any forced convention.

More practically, exploring what it means to have a voice and be an individual may require a rethinking of what kids study too. This is in part why you may be hearing more about the power of interdisciplinary study, where kids consider a far larger number of subjects as they move along toward college. So next we'll take a look at what subjects and majors can also support this journey toward individuality and in developing stronger EQ skills too.

Renewed Celebration of Interdisciplinary Study

I was talking to a parent recently who studied philosophy in college. We laughed about how in the 90s philosophy was dismissed along with its Liberal Arts peer subjects as impractical and the quickest forced route to a graduate school program. Yet today it's as in demand, and arguably as relevant, as computer science.

What changed? Well, for one, AI is an ethical puzzle that will never be fully "solved," and second, all the data we just covered about the softer, more "philosophical" skills required to get a job in the years ahead.

It's a win for a more diverse and "interdisciplinary" course of study, which at any time in history you can argue would have been the right choice. But now gone are the days that a child is a "math" person or an "artsy" student. Today, children can and should be all of these things and all at once. And not just to be personally resilient but to ensure a career is waiting when they graduate.

Opportunity to Academically Connect the Dots

Interdisciplinary programs are just as they sound—a deliberate mix of subjects like philosophy, science, math, and technology. Top schools such as Harvard, MIT, and Stanford already offer, or in some cases require, that students taking courses such as computer science also enroll in philosophy or ethics. For instance, Stanford's "Ethics, Society, and Technology Initiatives" allows engineers to also consider the ethical issues of the new systems they are creating.[7]

Universities such as Oxford and Georgetown also offer "AI and Society" courses where students study both machine learning and sociology. In many cases, schools are also creating majors that combine the fields. At many elite institutions, most notably Oxford University, the "Philosophy, Politics, and Economics" (PPE) major is one of the most popular and it is flexible enough for students to work in other related subjects to the mix.[8]

That said, on the side of academia, the practical implementation of this type of education hasn't been as smooth sailing as the popularity of the approach with students would suggest. Historically, these research headwinds are coming from the top thanks to an environment not fully incentivizing this type of structure.[9] But as AI further infiltrates the work that researchers are doing and their day-to-day experiences as academics, the opportunity to work more closely between departments is growing.

On a practical level, when it comes to our kids, this is about thinking more flexibly about solving the problems of the future. The ultimate goal is to train leaders who understand how AI systems impact issues like inequality, democracy, society, and culture and to approach decisions with a more diverse and creative lens. As we've just discussed as well, it's also about sharpening our EQ skills, which come with exposure to a much broader set of disciplines.

The Beauty of Academic Pursuit as a Puzzle

The traditional model of education—where students mostly pick a lane as they move into college and stay in it—made sense when careers were more predictable and technology changed the landscape slowly. But we're now living in an era where, to see the bigger picture, every expert must contribute their piece of the puzzle, and only when each piece is linked together do the answers become clear.

Think about what's happening in fields that seem completely unrelated to technology. Art historians are using AI to authenticate paintings and detect forgeries, but they need to understand both the algorithms and the cultural context of the artwork. Environmental scientists are analyzing climate data with machine learning tools, but interpreting that data requires understanding policy, economics, and human behavior. Even fields like social work are using AI-assisted case management systems, creating a need for professionals who understand both human psychology and algorithmic bias, data privacy, and cybersecurity.

These problems require different approaches to come together and complete the puzzle. This isn't just about career preparation either, but about citizenship too. As AI becomes more integrated into our daily lives, we need people who can ask the right questions about its use. Does this hiring algorithm perpetuate racial bias? Should this AI tutoring system have access to children's emotional data? How do we balance the efficiency gains of automated decision-making with the need for human accountability?

These questions require technical literacy, humanistic wisdom, and bravery. But to solve these issues first requires groups of people with diverse enough perspectives and varied expertise to come together, where from there they can start to look at recommendations and solutions.

How to Support Flexible Thinking and Pursuits

While the topic of interdisciplinary (or even cross-disciplinary) work is often talked about in the context of college majors, the spirit of what we're focusing on here can be accomplished at any age and in any environment. Put simply, it's an approach and a mindset, not a prescribed path.

For instance, a high school student might focus on STEM courses but then take electives in the arts. Or they might dive into the arts through extracurricular activities while maintaining a biomedical focus in their classes. Even just sampling a much wider variety of courses and discovering new ways to think about the past and future can connect the dots and strengthen the way children think.

When it does come to choosing a college, it might make sense to prioritize schools that encourage this type of interdisciplinary learning—formally or informally. The school doesn't need to offer fancy-sounding programs or the newest integrated degrees—that's often just marketing. It's about flexibility and the opportunity to mix subjects and majors.

This can look like the pursuit of double majors or schools that have foundational courses that every student must take that are more diverse in scope. Some of the most innovative colleges and programs I've seen aren't branded as "interdisciplinary" at all—they're just places where curious students can explore different connections between fields and are guided by supportive faculty who encourage boundary-crossing.

Families can model this mindset too. When you're reading the news together, point out how stories connect to multiple fields. An article about global AI data centers and their water demands involves science, politics, economics, and ethics. News about social media regulation touches on technology, psychology, law, and philosophy. Help your children see that the most important issues in our world require multiple perspectives to understand fully.

Building an Interdisciplinary Mindset at Home

Another way to encourage this kind of thinking is through creative questions in our daily conversations. Try asking "what if?" questions that bridge ideas. Like, "What if we designed cities the way we design video games?" Or even, "What if we applied principles from biology to computer programming?" And "What if we used theatrical techniques to teach history?" These kinds of questions open up the type of possibilities that single-discipline thinking might miss.

The goal isn't necessarily to create Renaissance scholars (though maybe it is!) but to help children understand that the most interesting problems and meaningful careers exist at the intersections of fields. In an AI-enhanced world, we need humans who can think across domains, ask ethical questions, and translate between different ways of understanding complex issues.

The companies and organizations that will thrive in the coming decades will be those that can integrate technical capability with human insight, that can innovate responsibly, and that can navigate the complex ethical terrain that AI creates. They'll need employees who can code and also understand user psychology, who can analyze data and also consider its social implications, who can design efficient systems and also ensure they serve future best interests.

When we encourage interdisciplinary thinking, we're not

just preparing our kids for future careers. We're helping them become the kind of people who can hold complexity with comfort, see nuance in debates, and approach problems with both analytical rigor and human wisdom. In a world increasingly shaped by AI, this type of intellectual curiosity might be the most important gift we can give to our kids.

Key Chapter Takeaways

If you started this chapter not having really considered the creation of AI-enhanced learning routines at home or in tackling the softer issues around EQ, I hope you're ending the chapter thinking it all makes perfect sense. I'd also like to congratulate you on getting this far overall in the book. It's difficult to consider all the risks that AI innovation brings and still forge on to consider the opportunities with optimism.

Successfully supporting your child's learning in an AI world isn't about becoming a technology expert or creating perfect systems from day one. It's about finding your family's personal groove and applying the same wisdom and solid instincts you've always used, just with some new tools and ideas in the mix.

You Already Know More Than You Think

The homework struggles, learning victories, and academic challenges you've navigated with your children have prepared you well for this moment—much more than you might realize. When you help your child choose between different AI platforms, you're using the same careful consideration you've always applied to selecting educational resources. When you set family rules about AI use, you're extending the values and expectations you've already established around honesty, effort, and learning. When you're having more probing or big-picture conversations, they're the same as you would have about any important family topic.

The strategies we've discussed throughout this chapter build on principles of effective learning that haven't changed just because AI entered the picture. Whether your child is using AI to understand math concepts, brainstorm essay ideas, or practice language skills, the goal remains the same—AI should accelerate learning, not replace thinking. AI is a tool on this journey of learning; it's in no way the destination itself.

Our Human Skills Will Matter the Most

Perhaps the most important insight I hope you've gained from this chapter is that AI integration into our lives actually highlights the value of distinctly human capabilities. The emotional intelligence we discussed, such as the ability to understand, manage, and work effectively with others, will become increasingly precious as AI handles more technical tasks in the years ahead.

Belief flexibility allows children to update their thinking when presented with new evidence. Personal voice makes each child's perspective unique and provides the antidote to generative AI's homogenized and watered-down communications style. These are no longer just "soft skills," but essential capabilities for the 21st-century workforce.

When we help children develop ethical reasoning alongside AI literacy, we're preparing them to make the complex moral decisions that AI cannot navigate. When we foster their ability to advocate for themselves and their values, we're equipping kids to shape how AI is used in their future workplaces and communities rather than simply accepting whatever systems others create.

Families who approach AI with a strong emphasis on human development create children who can collaborate effectively with AI while maintaining their authentic voice and values. These children then become the ones who will guide AI toward outcomes that serve humanity's best interests rather than being guided by AI toward outcomes they never chose in the first place.

Starting Where You Are is Perfect

Creating effective learning routines with AI doesn't require implementing every strategy immediately or achieving some impossible standard of perfection. Pick one or two areas that feel most relevant to your family's current needs and experiment with those approaches first.

Maybe you start by having regular conversations about AI use, trying age-appropriate platform selection, or establishing better routines around homework assistance. The evaluation criteria we discussed, such as safety, educational value, and family values alignment, provide a framework that you can apply to any new AI tool that enters your child's world. You don't need to become an expert on every platform and technique; you just need to become confident in your ability to assess whether a tool serves your child's authentic learning and overall best interests.

Remember that interdisciplinary thinking, which combines technical knowledge with human insight or scientific training with ethical reasoning, doesn't require a dramatic educational overhaul. It starts with encouraging your children to make connections between different subjects, to ask questions that cross traditional boundaries, and to explore interests that haven't traditionally "gone together."

When your child is frustrated with a group project, you're already building emotional intelligence. When a kid explains their passion for skateboarding to a skeptical adult, they're developing communication and persuasive skills. A child who volunteers their time at a local organization is building empathy and cultural awareness. These everyday experiences become the foundation for the kind of human-centered leadership that AI-enhanced workplaces will desperately need.

Your Path Into an AI-Led World as a Family

Every family's path through AI integration will look different, and that's exactly as it should be. What matters isn't achieving some external standard of AI use, but creating approaches that align with your family's values and support your children's authentic learning and development.

The strategies you've developed through this chapter will continue evolving as your children grow and as technology advances. Trust your parenting instincts, stay connected to your children's experiences, and remember that you're building something much more important than perfect AI habits—you're raising thoughtful, adaptable learners who can thrive in whatever world they inherit.

You don't need to have all the answers about AI to support your children effectively. What you need is the willingness to learn alongside your kids, the confidence to apply your existing parenting wisdom to new situations, and permission to be wrong and adjust. This foundation you're building today will serve them—and the world they'll help create—for years to come.

Chapter 9
How We Can Build a Better Future Together

Recently, my friends Richard and Emma shared how their 14-year-old twins were excitedly talking about the robot in their science classroom and asked Emma what it might be like when these machines look like humans. At the same time, Richard was reviewing an email about cell phones in school, reading an article suggesting generative AI makes us "dumber," and talking to a friend about social media and political division.

"We really don't know which way to look right now," they said. "We find it so difficult to reconcile the idea that while people are fighting on social media, our kids are mentally preparing to welcome robots into their day-to-day lives."

Richard and Emma are wrestling with something we all need to find some comfort with—the ability to look near and far simultaneously. This is particularly true when the "far" is coming at us so quickly. For instance, specific to the robots Emma was talking to her twins about, experts estimate that there will be 1 billion of these "humanoids" on the market worldwide by 2050.[1] And while the technology isn't quite there yet, the industry is already valued at $6.8 billion, with

China leaping ahead of every other country in their development of these human-like machines.[2]

Humanoids won't just fill in with support to do labor-intensive work; they will also serve as social "companions" intended to interact with and provide emotional support to general consumers.[3] This makes talking to your kids about robots less science fiction and more a practical conversation to have right now.

This should also help you connect the dots on something we discussed earlier: Remember those AI companions already wreaking havoc on kids' (and adult) psyches? Now imagine them with physical forms that look and move like humans. Understanding where the humanoid industry is headed makes the stakes of those digital relationships even clearer.

Our Collective Journey to This Moment

We've spent this book examining the challenges that come with AI entering our families' lives and how to approach all these issues with confidence and independence. We've discussed privacy concerns, educational policies, the importance of human connection, and the skills our children will need to thrive in the workplace of tomorrow.

But there's one last side to this story that deserves our attention—the remarkable possibilities ahead of us and how we can prepare for it to come at us in fits and starts in the years ahead.

While we may have been caught off guard by ChatGPT, I hope you now see how it was really decades in the making. Society has had more than enough time to prepare for these big existential questions and wrestle with practical considerations —but we were never given a collective heads up. It won't be decades again before the next disruptive innovation arrives.

We're here now, and we need to have these conversations right away and continually going forward.

It's okay to admit that we're worried, but what I hope you've discovered throughout our journey together is that the fundamentals of raising a family and guiding children will stay constant. The beliefs and values you hold today, and even held yesterday, are still the best guide to preparing for what's ahead. Critical thinking, creativity, empathy, and genuine human connection don't become obsolete in an AI world; they are the superpowers that will allow us to stay in control of these machines.

What You'll Discover in This (Final) Chapter

In this final chapter, we'll focus on how you can take your newfound wisdom and think more critically and expansively about the big issues and opportunities that are coming next.

I don't know exactly what will happen in the years ahead, but as I've shared previously, I'm 100% sure that we have an opportunity to reset the terms of our collective engagement with the technology industry. We are no longer passive observers and consumers, but "shareholders" by nature of our data being used to build AI innovation. We get a metaphorical vote and say in what happens.

I'm going to argue in this chapter that AI innovation does not mean we become even more tightly tied to technology or that we hand over our thinking to machines completely. Innovation can give us back our independence. We have a chance to log off and fully reconsider our entire relationship with the digital realm in the embrace of AI innovation—as counterintuitive as that sounds.

So we'll consider if AI might actually free us to spend more time engaged in the human experiences that matter most, from unstructured play and big family dinners to cross-generational relationship building and face-to-face problem-solving.

We'll also discuss how it's up to us to decide if AI makes us smarter or whether it encourages intellectual laziness—as AI is not (and never will be) our master but simply a tool of discovery and efficiency. AI is and will continue to be built on our terms if we collectively recognize this fact and take control.

I'll share how recommitting to our local communities, engaging with neighbors, and making new friends will help build the policy advocacy and grassroots alliances we talked about earlier in Part Three. And finally, we'll take a serious look at those humanoids my friends Richard and Emma mentioned and the very practical—and deeply emotional, psychological, and existential—implications of living alongside machines that look like us.

It's easy to dismiss this topic as unrealistic "sci-fi" talk. But considering the rapid industry growth, we need to talk about it now. These conversations might feel strange, but they're necessary.

By the end of this chapter, I hope you'll see that AI innovation isn't a threat to our children's future but the rocket fuel to power their potential—if we choose to use it that way. You should feel hopeful about the future, not because it will be predictable but because it will be led by a generation we've prepared well. We are in control, we have the power, and I hope soon many more people will feel emboldened by this fact.

Freeing Ourselves from Digital Handcuffs

There are few moments in our day-to-day lives that don't include some responsibility or commitment with a digital component built in. We spend so much time online that it's difficult to believe that will ever change. In fact, a 2025 survey found that Americans spend an average of 10 hours a day online (and many families probably would estimate that number to be even higher).[4]

But what if I were to tell you that one of AI's most promising potentials is in helping us to spend our time online more intentionally—and therefore fewer hours plugged in?

Rather than endlessly scrolling or consuming content passively, we might use AI's efficiency and preciseness to get the job done, leaving us to move on to more meaningful activities. Even with the best intentions, sometimes it's a lot of work to get what we need online and not necessarily time well spent either. Often we merely end up deep in mundane admin life tasks or doing time-consuming research. We are also spending a shocking amount of time "doomscrolling" through our social media feeds—an average of 3.5 hours a day, according to a recent survey.[5]

What if we stopped? And what if AI could support those mundane tasks on our to-do lists and set us free? Certainly the idea of doing fewer tactical and rote activities and more of what we love is a good thing. In what ways could we use all those reclaimed hours?

For one, perhaps getting our kids offline and outside. Experts have been suggesting for some time that we should encourage our children to participate in many more hours of outdoor, unstructured play. NYU professor and author Jonathan Haidt has been for years an advocate for free play and in untethering kids from their devices. In his book *The Anxious Generation*, the author shares research illustrating how the rise of social media use with kids intersects with their declining mental health.[6] Haidt believes that we've allowed our always-on digital world to persist too long, and now a generation is suffering because of it.

If understanding the mental health issues of being tied to our devices isn't enough to get kids to voluntarily log off, losing their competitive edge in the work world should do the trick. Developing those EQ skills we talked about does not happen online.

Regaining the Freedom We Once Knew

Many Gen Xers remember those public service announcements that asked parents, "It's 10 p.m. Do you know where your children are?"[7] The fact that kids in the 70s and 80s had so much freedom that their parents needed this kind of reminder seems unbelievable today. Equally shocking is the idea that when we headed out, there was no way to track us and little ability to get in touch either.

Fast-forward to 2008, when writer Lenore Skenazy published a short op-ed about letting her 9-year-old take the NYC subway home alone.[8] The outrage was swift and intense. Even though the relatively short journey was well considered and organized, and at a time when NYC was objectively safe, the criticism Skenazy received was astounding. But the experience made her an even fiercer advocate for child independence, and today she runs "Let Grow" with academics and leaders, including professor Jonathan Haidt.[6] The organization's mission is to promote child independence, and our interest in the topic today shows that Skenazy may have been on to something all those years ago.

We've become so dependent on our devices that we've chosen the comfort of surveillance over the opportunity for freedom and personal responsibility that comes from turning it all off. And we've found ourselves in a place where digital socialization is the norm versus the messier work of interpersonal conflict and negotiation. When did we decide to trade our freedom for these digital handcuffs?

All is not lost, though, as technology overdependence is not a requirement for benefiting from future innovation. We don't spend hours online because it catapults us into great opportunity and discovery, but because our mindless scrolling and clicking is addictive (and profitable for Big Tech). Embracing our "free range" roots again is more than possible; it's imperative to succeed and survive. The generation raising many of the kids today experienced this freedom and just now needs to tap into this memory to make a lasting change.

Logging Off as a Competitive Advantage

As AI systems both introduce efficiency and create demand for the "softer" skills in life as we've covered, we'll start to see less time online as a measurable competitive advantage. For instance, the AI-first, US-based Alpha School has been in the news recently as it expands into new cities.[9] The school takes a radical approach to education: students spend just two hours a day in AI-powered instruction and then dedicate the rest of their time to "life skills" such as bike riding, public speaking, sailing, entrepreneurship, and even organizing their own playdates.

The novelty of Alpha School teaching kids via an AI system for just two hours a day tends to get the most attention from the media. But what strikes me equally is the fact that roughly three quarters of Alpha School students' school days are spent offline and in physical, interpersonal, and outdoor activities.

These activities are not just pleasant breaks from screen time either; they are a chance to rewire the brain and hone the important EQ skills I've mentioned over and over again. In a world where AI can handle information processing, the kids who can think creatively, solve problems with their hands, and navigate uncertainty will undoubtedly stand out. And clearly, Alpha School and similar schools and programs understand this fact.

Nature's Incredible Impact on the Brain

Unstructured play, especially outdoors, has always been crucial for child development. When children build elaborate imaginary worlds, negotiate rules for backyard games, or spend hours exploring tide pools, they're developing creativity and problem-solving skills that no AI interaction can provide.

The physicality of outdoor play also provides sensory experiences that support healthy brain development and stress regulation. Climbing trees, running through sprinklers, or simply lying on the grass watching clouds swirl about requires the kind of embodied presence that balances screen-based learning.

What might come as a surprise, though, is that nature has also been proven to help us think better. One of my favorite books of the past couple of years is *The Extended Mind*.[10] Author Annie Murphy Paul cites research on how external stimuli, from groups of people to sounds and other aspects of our personal environment, can impact our thinking. And most intriguingly, that nature, such as grass, trees, and other foliage and outdoor elements, contain tiny visual patterns that aren't easily replicated by humans. These difficult-to-detect but powerful patterns can have a measurable impact on our cognition. By just looking out a window onto a nature-filled landscape, we can actually think better.

The point is that nature, the outdoors, and human interaction will strengthen our brains in ways superior to machines, and we need that advantage to work with and not for machines. And just like a lot of what I've shared in this book, these new realities aren't nice suggestions for future consideration but imperatives to ensuring our success and survival.

Rethinking the Hardware We Will Want and Need

The next wave of technological development is likely to provide us with the extra support to rethink how we get away, unshackled. Augmented Reality (AR) and Virtual Reality (VR) technologies, for instance, where devices such as glasses are thoughtfully integrated with AI, could actually encourage families to engage more fully with their surroundings.

Picture AR systems that make family hikes feel like treasure hunts, or where kids might point their phones at unusual rocks and instantly learn they're looking at a 300-million-year-old fossil. A family camping trip could become an opportunity to learn about astronomy, geology, or ecology in real-time, with AI providing personalized explanations based on a child's age and interests.

VR experiences could transport families to historical moments or distant places, then inspire them to explore their own local history with new appreciation. Instead of pulling kids away from nature, or in the case of VR applications, history, these tools could make every outdoor or new experience feel like a personal discovery. It can feel like something they want to chase and learn more about.

Of course, like anything we've covered in the pages before, it's also easy to see where it all can go wrong. Especially when talking about VR and the possibility of getting sucked into worlds that do not exist and never tapping back into our own (we've seen this movie before). But it doesn't make the possibility any less compelling; it just highlights how important smart, ethically focused leadership skills will be to designing every innovation going forward.

A New Vision of the Future Coming Into Focus

We are moving toward a future where AI efficiency could actually restore some unstructured childhood experiences that previous generations took for granted—if we allow it to. If we do it right, children could have more time for free exploration, physical play, and authentic human connection—the experiences that support healthy development and give us back our freedom.

The goal isn't rejecting AI or returning to a pre-digital world, but thoughtfully integrating AI in ways that enhance rather than diminish the experiences that make childhood rich and meaningful. Children who grow up with this balanced approach will be prepared to not just use AI more effectively but also to aid human flourishing—making us smarter, more successful, and hopefully happier, in the end.

AI Will Make Us Smarter if We Let It

There's a persistent worry floating around that generative AI may make our children lazy thinkers, or more dramatically, "dumb." The concern goes something like this—if kids can get instant answers to any question, solve math problems without doing the work, or have essays written for them, why would they bother developing reasoning skills? Or even do the work at all?

But I hope that in reading this book, you might now realize that we are far more able to decide how AI innovation is used and what the outcome should be than we are allowing the current narrative to suggest.

Many academics and researchers also agree that AI innovation is what we make of it—and that can mean making it something bad too. At the Stanford AI+Education Summit in 2023, researchers noted that AI not only offers new opportunities to teach kids, it also carries the risk of automating bad ways of teaching.[11] We must do the work to train and empower teachers and decide how this should all unfold. AI reflects our essence—the good and the bad—and we can't lose sight of that fact as well.

For adults who are already mature, educated, and critical thinkers, the access that AI systems provide to create quick research, more in-depth insights, and support for time-consuming tasks means the chances of getting "smarter" are excellent. Research is giving weight to this claim as well, showing that AI-human collaboration is most effective at tasks that humans already do quite well—so just enhancing this expertise in the end.[12]

What's harder to say with confidence is whether AI systems can make someone "smart" in the first place—and that's where we need to return to schools, prepare teachers, and think carefully about how AI is used in education. We can't just assume that because AI is efficient and sounds authoritative, it belongs in every aspect of learning. Just because a tool works well doesn't mean it's the right tool for every job. We must be thoughtful and deliberate about where and how we use AI innovation.

The Too Convenient, Unhelpful Villain Narrative

When we heap too much scorn or praise on AI innovation, we also do damage to our understanding and ultimately use of AI.

We are looking too hard for the good guys or the bad guys and not peering as closely into ourselves. This can also create a lot of noise and content that obscure the truth and our ability to consider these big questions.

Ironically, as we discussed in Chapter One, this type of hype cycle without context is precisely what damaged AI investment and advancement decades ago. So we should pay particular heed to these past lessons and not allow dramatic narratives to take over.

For instance, many media outlets ran dramatic headlines in mid-2025 suggesting that MIT researchers had found that "ChatGPT makes us dumb."[13] The actual research, led by Nataliya Kosmyna at MIT Media Lab, had looked at the impact of generative AI on a user's brain function and simply found a reduction in "cognitive load" for those using generative AI.[14] And no, that's not the same as making one "dumb," it's in fact a misrepresentation of the researcher's work (something she was forced to take to LinkedIn to explain).[15] Unfortunately, as is often the case, the damage had been done, with little chance that readers would later seek more detail or that an outlet would make a correction.

Blaming a technology system for poor educational outcomes is far too convenient a way to explain away AI's complexities. We don't seem to want to accept that it's complicated or that we have more agency over it all than we want to admit. Even the "AI makes us dumber" argument is nonsensical in that it assumes that intelligence is a finite resource that gets depleted when we delegate cognitive tasks to machines. We're all too smart not to see the fallacy in these arguments.

History Can Be the Guide to Finding Our Way

There are many examples in the past of times when a new tool or technology was approached with irrational skepticism. We talked earlier in the book about the Internet's introduction to the classroom. But we've seen these questions going even farther back with the calculator or word processor.

In the case of the calculator, the history of its acceptance in classroom settings was a much longer journey than many of us might realize.[16] It wasn't until late into the 1980s that educators even began to allow calculators in classrooms, and not until the graphing calculator became available did it become a universally accepted staple. We ultimately found that calculators did not damage kids' ability to learn math.

There are also parallels in the use of word processors. When students use word processors with spell check and grammar assistance, they don't become worse writers. You could argue that they become better writers because they can focus their cognitive energy on developing ideas, organizing arguments, and crafting compelling narratives rather than getting bogged down in the minutia of spelling.

This is not to dismiss any of the valid concerns about generative AI in the classroom, but when we worry about its impact on education or longer-term intelligence, we have historic precedent to consider. We also have far more control over how AI is used (as we did with these other devices in the past), and the power is fully within our hands.

AI's Potential as a Thinking Partner

One way to address concerns about generative AI's impact on cognitive development is to ensure the technology is used as a thinking partner—and that we talk about it in this way. It should augment learning, not outsource it.

When children use AI as a replacement for thinking, they're bypassing the cognitive work in a way that does not strengthen their capabilities. Asking ChatGPT to write an essay, solve math homework, or answer reading comprehension questions falls into this category. The immediate task is done, but it's definitely not the way to get smarter.

But when a child uses AI as a thinking partner, it's something different—it can enhance learning and help kids develop higher-order thinking skills. It's not about asking generative AI to do the work, but to help with the work. It is in the delivery and intention, and, again, that puts much of the control in our hands as a result.

If we let it, AI can allow for "productive" struggles, where we ask it to challenge us (or families program the system to do this). Good learning happens when students wrestle with ideas just beyond their current understanding. AI can provide the scaffolding that makes this kind of productive struggle possible. Instead of getting stuck on basic definitions or background information that causes them to get distracted, their mental energy can stay focused on analysis, synthesis, and creative problem-solving instead.

The point is, AI can actually make a child feel that the work is "harder" and more time-consuming because they are being forced to think harder. And that's when you know AI systems are augmenting, not outsourcing the work of thinking.

The Real Risk in Not Giving AI a Chance

The real risk isn't that AI will make our children lazier thinkers. The real risk is that we'll use AI poorly and miss the opportunity to amplify our intelligence. This is especially true if others decide to run with it, creating a wider disparity in understanding and core ability.

When we frame AI primarily as a threat to intellectual development, we discourage the kind of thoughtful experimentation that leads to breakthrough insights about how humans and machines can work together effectively. The most successful students in an AI-enhanced world won't be those who avoid AI tools, but those who become skilled at using them and who explore the different ways AI can amplify their already well-formed thoughts.

Imagine children who grow up comfortable asking AI to help them understand complex scientific concepts, then use that understanding to design their own experiments. Picture teenagers who collaborate with AI to research social issues they care about, then channel that knowledge into advocacy work that makes a real difference. Think about families who use AI as a starting point for deeper conversations about ethics, creativity, and what it means to be human.

That's what it means to use AI to get smarter. This is what's possible when we approach AI with curiosity rather than fear. When we teach our children to see these tools as extensions of their own capabilities rather than replacements for their thinking. When we help them understand that the goal isn't to compete with AI but to dance with it.

That's not lazy thinking. That's the future of human intelligence, amplified and enhanced by artificial intelligence. And the only way we get there is by spreading the word, working together, and building alliances in our communities and elsewhere.

Making Friends to Build Alliances for the Future

It's remarkable when you think how easily we've given up our worldview to those with the algorithmic power to shape it. We also contribute by willingly defining the people around us like they are data points on a spreadsheet—"left-wing in this column," "right-wing in that quadrant." Social media platforms, in particular, have mastered the art of forcing us into tidy packets of categorization, and all for the purposes of generating ad revenue and the bigger jackpot of data to train and build AI.

Media have also increasingly followed suit, with the number of more dramatic and hyperbolic headlines increasing year-over-year as media companies seek to protect increasingly their fragile revenue at our expense. Researchers have noted that even traditional "mainstream" outlets, such as *The New York Times* or *The Guardian*, have exhibited a sharp increase over the years in headlines that would be considered "clickbait."[17]

At the same time, the local news that used to inform and connect us has been disappearing at a steady rate. According to a study by my alma mater, Northwestern University's Medill School of Journalism, since 2005, more than 3,200 print newspapers have shut their doors, and today newspapers are closing at a rate of two per week.[18] This further disconnects us from the world in our vicinity, where the important work of refining those much-needed "human skills" and building alliances takes place.

There are many reasons we must focus on in-person connection in an AI-led world. And most urgently, I'd suggest the top ones are to create alliances for policy and other grass-roots efforts, address digital isolation and loneliness, and ensure widespread digital literacy so we all can benefit. All of these things need us to recommit to reconnecting with one another.

Importance of Focusing on Issues, Not Tribes

When you strip away the political noise and focus on what's actually happening in people's daily lives, you discover that people across any given perceived political spectrum share remarkably similar concerns about kids and the future. And that will be even more apparent as the months and years go by, when it's clear that many of these issues don't have an attached political party.

Think about what unites us right now—concerns about screen time and social media's impact on mental health. Worries about whether our kids are developing adequate social skills and questions about academic integrity when generative AI is part of the picture. Anxiety about whether kids will be fully prepared for the workplace and what credentials and skills will be needed in the years ahead.

These aren't political issues; they are human issues. The same pattern shows up when we talk about AI regulation. While politicians are increasingly working across party lines on issues such as data privacy and AI regulation, families are doing the same. Technology issues affecting our kids show far less division than almost any other topic dominating our national conversation right now.

What's particularly interesting is how AI might actually take this one step farther for us and fully remove these divisions completely. In fact, a YouGov poll in April 2025 found consistent concerns about AI regardless of party affiliation.[19] Republicans, Democrats, and Independents were aligned on all questions, from whether AI will make Americans more productive and less lonely to whether the industry should be better regulated. The latter was particularly surprising, since regulation is often a partisan issue. And even when race, age, gender, ideology, or education were factored in, Americans remained aligned.

This is remarkable moment that deserves more of our attention. We live in a time when we seem to barely agree on anything, yet here's an issue where common ground exists across every demographic divide. If we don't pay attention to this rare alignment, we miss a critical opportunity to exert our collective power when it actually matters most.

Think about what we could accomplish if millions of families, regardless of background or political beliefs, spoke with one voice about protecting children's data, ensuring ethical AI development, or demanding transparency from tech companies. That's not a coalition that can be easily dismissed or divided by the usual political tactics. That's the kind of unified pressure that actually changes policy and shapes how technology develops. But only if we recognize the opportunity in front of us and act on it together.

Everyone Needs More Friends Right Now

We talked in earlier chapters about the impact of AI technology on the loneliness epidemic. But core to this is the skill of making friends, and we are woefully out of practice. According to the American Perspectives Survey survey, in 1990 only 3% of Americans said they didn't have any friends, and by 2021 that number had risen 12%.[20] While we may have once considered social media to be a way to better connect, research has actually found that instead of making people feel connected, social media can lead to a sense of isolation.[21]

The result is that more than ever before, people are without the human, in-person connection they need. And as we covered earlier, a lonely, friendless society is not a healthy one (literally) either, nor is it one where we can solve any issues related to AI.

Friendship-making is a skill that needs to be improved with practice, just like ballet, knitting, or playing the piano. But we need to do the work to get there. As adults, we can do a better job of modeling friendship-making behavior, particularly in fostering intergenerational connections.

When people of different generations connect, children learn incredibly important capabilities like the natural rhythm of conversation. They figure out how to listen, when to jump into a discussion, and how to read facial expressions and body language. They practice the give-and-take of group conversations, learn to disagree respectfully, and discover that other people have different perspectives actually worth considering.

Most importantly, if we are going to tackle difficult, emotional, and even larger existential issues, we need one another. And our kids require the type of authentic, solid relationships that give them the necessary grounding as they face the roller coaster ups and downs of the future.

We Need Widespread Digital Literacy

It goes without saying that digital literacy across every level of society is essential—and without it, critical consideration of AI and its impact is almost impossible. But despite our enthusiasm for all things digital, we have not, in fact, prioritized literacy. And not only is literacy best acquired through relationships and encouragement, its absence makes building alliances and doing important grassroots work impossible.

According to a survey conducted by Pew Research Center in 2023, a majority of the Americans polled answered just five of nine digital literacy questions correctly. [22]Questions included topics such as cybersecurity best practices, facts about Big Tech companies, and federal privacy laws. For instance, true or false statements like "websites are legally prohibited from collecting data from kids under the age of 13," or what a deepfake is and that LLMs learn from data pulled from the Internet, were tough for some to answer correctly. And only 4% of respondents could answer all nine questions correctly.

The reality is that AI policy and regulation will be shaped by those who understand it best and engage most actively. If we want family-friendly AI policies, we need families who understand the details. This means helping our neighbors, extended family, and community members develop the same literacy we have for any category of need in our lives, from our health to what we eat and how we take care of our homes.

Digital and AI literacy require patience, encouragement, and understanding to teach, and it's best achieved between friends. Not only that, but it's also critical we bring everyone along. Without literacy skills, an elderly couple is more likely to be duped by deepfakes. Parents on a PTA might have documents stored in a shared Google Drive compromised if they don't have a solid understanding of cybersecurity best practices. And schools might find themselves at odds with families if no one is having the same conversation about generative AI.

We simply can't function as a society going forward if everyone isn't moving in the same direction with the same basic understanding. We must build a community of informed citizens who can ask good questions, spot potential problems, and work together toward solutions that serve everyone's best interests. It's like a community watch group, but for AI and the future advancements that come with it.

We'll need these community connections and the shared knowledge they bring even more as AI takes its next evolutionary step from our screens into our physical world.

Preparing for Shocks (Even Good Ones) Ahead

The launch of commercially available generative AI technology is only the beginning. What comes next will bring wonder, invention, and excitement, but also disruption, disagreement, and confusion. Our way of life will be tested, and we need to prepare for that reality now.

There will be great discoveries and incredible solutions to societal problems. There will also be upheaval and sci-fi-like scenarios coming to life faster than we expect. But here is something important to remember: much of what we'll struggle with —whether amazing or problematic—has less to do with the technology itself and more to do with how it makes us feel.

Take humanoid robots. As the market for robots that resemble humans grows, we may not actually find them all that useful in our daily lives. But their very existence will spark confusion, unease, and for some, genuine emotional distress. Similarly, we'll celebrate breakthrough scientific discoveries enabled by AI, but those same breakthroughs may eliminate entire fields of work overnight. The wins and losses will arrive simultaneously, and we'll need to hold both realities at once.

This is why emotional and psychological preparation matters as much as understanding the technology itself. We can't predict every development, but we can prepare ourselves to respond thoughtfully rather than reactively. We can build the resilience to adapt when things change suddenly. We can teach our children to navigate uncertainty without panic.

By now, you should feel more equipped than when you started this book. You have frameworks for thinking through these challenges, language for discussing them with your family, and an understanding that you're not facing any of this alone. Whatever comes next, you can approach it with clearer eyes and steadier footing. That's not just preparation—that's power. And so what can we expect next? Here are a few areas to pay attention to:

An Emotional Reckoning Coming via Robots

As I mentioned at the start of this chapter, the market for robots that "look like" humans (two arms, two legs, etc.) is already nearing $7 billion in value and growing steadily. If you think about these as being the natural "next step" after chatbots that sound human, it seems to make sense. Companies like Boston Dynamics, Tesla, and dozens of startups are developing robots designed to work alongside humans in homes, offices, and public spaces.

The most recent iterations of robots are clunky, obviously mechanical, and with very narrow capabilities. But the goal is, undoubtedly, to develop systems that will look increasingly human-like, respond to natural language, and perform complex tasks that require physical dexterity and social awareness.

But as all of this is building, we'll start to hear more bizarre stories and fantastical human-machine interactions, like people hoping to "marry" their robot companion. Don't get distracted though; there is something deeper and more serious to consider than just the most intentionally shocking headlines.

Even if we are a long way off from the movie and TV version of sophisticated, eerily compelling robots, just the idea of these will be emotionally hard for most of us to deal with—and particularly our kids. For children who are already forming relationships with AI chatbots, the introduction of physical AI companions starts to raise serious questions that we might not be fully prepared to tackle right now. But we do need to start thinking about the types of human-humanoid relationships that will be reasonable to expect in the future and what conse-quences there will be. Like how emotional entanglement with humanoids might stunt humans emotionally or start to chip away at the fabric of society.

If we're already letting one-dimensional companions get the best of us, what happens when they "look" human? Is there a tipping point here? We must be willing to talk about it without worrying we sound a bit crazy. In fact, I've been talking and writing about the human-robot future dynamic for years and have received my fair share of strange looks as a result.

This isn't necessarily some dystopian future we are talking about either, but another very practical reason for families to be intentional about their AI use. And work harder and faster at better cultivating distinctly human experiences and relationships with bonds that can't be broken by machines.

The good news is that this topic should make for some interesting and lively family conversations, and we should embrace the opportunity to use its absurdity to dig into family values, beliefs, and any concerns our kids already have. We should hear from our kids too about what they think makes relationships meaningful. Help them to think through what they value most about their friendships and family connections. These discussions will help them navigate a future that is starting to stack up in its complexity

AI's Exciting Potential for Scientific Discovery

We often get so caught up in the story of the moment or the thing that draws us onto social media (or the absurd, as we've seen here), but that can cause us to miss the actual incredible wins that AI has already brought to the world.

The first Nobel Prize for an AI-based innovation was awarded in 2024. John Jumper and Demis Hassabis at Google DeepMind in London were the duo honored.[23] The scientists achieved a breakthrough that had previously eluded researchers for decades: understanding how proteins fold into their complex shapes. Developing an AI system called Alpha-Fold, the team created an AI model with the ability to predict the structure of almost every known protein.[24]

The discovery's significance touches everything from drug discovery to solving environmental issues like how to break down plastics. It's probably one of the biggest AI news stories of the past two years that many people did not hear about (although there is a great documentary worth watching on the subject).[25]

AI has also had an impact on the way scientists and researchers work overall. Information overload can be an irritating and unnecessary obstacle to the important work that they do. For instance, every year, an overwhelming number of scientific papers are published—far too many for any single researcher to read or keep track of, even with relatively sophisticated search platforms. But now AI-powered tools can analyze mountains of data and give scientists the chance to stay focused on finding meaningful research connections rather than spending hours poring through research.[26]

There are many examples like this, and the implications for how quickly we can now reach groundbreaking insights are one of the more powerful upsides to AI that we can't overlook.

Using AI to Help Solve Social Challenges

Perhaps most striking is AI's potential to accelerate solutions to stubborn social problems. In healthcare, algorithms are already helping hospitals match organ donors with recipients more efficiently, potentially saving thousands of lives each year.[27] In education, predictive models can flag students at risk of dropping out so teachers can intervene earlier.

AI also plays a role in global challenges, with conservation groups using machine learning to predict poaching activity and protect endangered species.[28] Farmers are now adopting AI-driven sensors to reduce water use while improving crop yields, making agriculture more sustainable in a changing climate.[29] Social workers are experimenting with AI tools that help families access housing, food, or counseling more quickly—sometimes even preventing crises before they unfold.[30] The examples are endless.

NASA and other space agencies rely on AI to operate rovers and satellites in environments too far away for real-time human control. As systems grow more autonomous, they'll help us search for life on distant planets, manage long-duration space missions, and analyze astronomical data on a scale humans never could. The discoveries made by these AI explorers could fundamentally reshape how our children understand humanity's place in the universe.

Preparing for a World of New Possibilities

The real challenge isn't whether these tools will exist—it's how we choose to use them and the problems society chooses to focus on. You are now even more aware than you were likely before of the risks that AI carries alongside its promise. We talked about how algorithms can reflect biases in the data they're trained on, sometimes producing unfair outcomes in hiring, policing, or lending. Privacy concerns are real, particularly when sensitive health or education data is involved. And there's the risk of over-reliance—technology should empower children, not replace human relationships or creativity.

That's why preparing our children for this world requires more than teaching them how to "use AI." It means teaching them to think critically about technology, the future, how we address problems, and who is "in charge."

We want our kids to ask questions about who benefits from a given technology and who might be left out. We need them to consider the values of an AI model's creators and work to develop healthy skepticism about marketing and efforts to sway us. We need them to be discerning but also keep these types of meaningful innovation initiatives going.

The future our children inherit won't be perfect, and of course no technology can eliminate all suffering or solve every problem. But it could be a world where we make significant advancements in curing preventable disease, where scientific discovery accelerates in ways where people live better and have more opportunities, and where technology amplifies human potential rather than replacing it. We all should want that.

Our kids will build on the choices we make today—and that's a weighty responsibility for all of us. But if we all approach AI with curiosity rather than fear and teach our children to see technology as a tool for a better tomorrow, we may actually be able to achieve all of it.

Key Chapter Takeaways

If you started this chapter feeling full of the skepticism and curiosity that I've advocated throughout the book and now feel more confident in your responsibility, then it's been a success. These are the types of topics, opportunities, and issues that should make us feel emotional, hopeful, and powerful. It should also be clear now the weight we bear for decisions that shape the future and how none of us is exempt from carrying these responsibilities.

Navigating the rocky waters ahead is something we approach with everything we have. We need to fight for knowledge, for our humanity, for children, for the environment, and for our sanity. We must resist being psychologically manipulated and swayed. We need to take every opportunity that AI presents to us and believe that we can live better lives.

It's critical now too that we see each side and each angle of any story, problem, assertion, or idea. The truth is complex. And we are not on different teams related to our political beliefs, our cultural identity, our religion, or nationalities. We are only team humans when it comes to a world shared with machines.

Embracing a "Shareholder" Mindset

A seismic shift could be happening now if we embrace transforming from passive consumers of technology to active stakeholders of AI technology. Our data, our conversations, our creativity—all of this feeds the AI systems that are reshaping the world. This means we are financing someone else's success with our every being, and it's not clear that we are getting as much as we should in return.

We're not just buying products anymore; we're contributing to the intelligence that will shape the future. That gives us both great responsibility and enormous power. But only if we allow for it. And in doing that, we must rethink how we work with commercial interests and technology companies. There is no more "but we aren't smart, savvy, rich, educated, informed" enough. If companies want our data for AI models, or our land for data centers, and our participation in favorable policies, we should be treated like business partners, not passive consumers.

How do you try on this "shareholder mindset" and take it for a spin? Well, first, stop doomscrolling and getting pulled into endless, meaningless, increasingly bot-led social media arguments. Choose platforms and tools that actually serve your family's goals or needs. Spend time when it's beneficial, and turn things off when you are done. Say "no" to surveillance of any form, and think about what you want from tomorrow and are ready to ask of the corporations designing new technology. Be the "boss" and train our kids to think that way too.

Embracing the Freedom to be Fully Human

One of the biggest misconceptions about AI is that it will diminish the value of human capabilities and somehow make us "less smart." But I hope you see that this is a choice. The breakthrough with AI isn't just about making machines work harder for us—it's about freeing ourselves from the digital dependencies that have consumed so much of our time and mental energy.

When AI can handle research, organize information, and streamline routine tasks, we get back precious hours to spend on what matters most—unstructured play with our children, meaningful conversations over family dinners, face-to-face problem-solving with neighbors, and the kind of cross-generational relationships that build real wisdom.

These aren't just nice-to-have experiences. They're where children develop the human capabilities that become superpowers in an AI-led world. AI gives us permission to prioritize these experiences without feeling like we're falling behind.

There's a sense in all of this that we can achieve something of a "human reset." We're not lurching ahead to a cold, machine-driven future. Nor are we relegating ourselves to years of continuing to bury our heads in devices. We can break free from all of it. Our digital lives, particularly with their over-reliance on social media, have become incredibly passive and reactive. AI gives us the tools to step away from this reactive digital existence, and we should take it.

Bracing for Machines That Look Like Us

The introduction of humanoid robots into our daily lives isn't science fiction anymore. With 1 billion humanoids expected by 2050, our children will grow up alongside machines that look and interact like humans. This requires both practical and emotional preparation.

Practically, we need to help our children understand the capabilities and limitations of these systems, just as we've done with AI chatbots. Emotionally, we need to establish boundaries about what kinds of tasks and interactions are appropriate. When machines look human, our children's brains will naturally want to form human-like attachments.

We need to start conversations now about what makes relationships meaningful, what they value most in their friendships and family connections, and how to maintain authentic human bonds in a world where artificial companionship will be increasingly sophisticated. These aren't distant concerns— they're conversations we need to have at the dinner table today.

Millions of Voices, One Important Conversation

The future isn't predetermined by AI development. Your children will play active roles in shaping it through the choices they make, the problems they choose to tackle, and the values they bring to their work and relationships. By helping them develop both AI literacy and deep human wisdom, you're preparing them not just to succeed in an AI world but to lead in creating the kind of future we all want to live in.

This is why your family should feel hopeful about AI. Not

because the technology guarantees good outcomes, but because you're raising children who have the judgment to guide the technology toward human flourishing.

The conversation about AI and families doesn't end with this book. If I've been successful in getting you to this point, it's because you are ready to go farther. Theses important discussions with your family are just beginning—and now you can take them and run. Keep talking at the dinner table, in school, at the store, and in those countless other moments during your day-to-day life.

The answer to where we go from here is simple—we need to talk about it. Millions of people talking about what AI innovation already means to their work, lives, and those of their children, loved ones, and neighbors could change *everything* about what's next.

We don't have to agree about AI technology, its use, or the future of innovation. In fact, it's even better if we don't all agree, as that is the type of friction that often produces the best solutions. But there is simply no AI without all of us on board. And the sooner we spread the word related to this fact, the better for humankind—and not the least of which the children that will lead us into tomorrow.

Afterword: How I Used AI

Did I use AI to write this book? You bet I did. And honestly, would you trust a book about thinking through and using AI if the technology didn't play a major role in the book's development?

I leaned on AI heavily to do everything from refine my outline and source research to unpack the ins and outs of self-publishing, nail tedious details like citation formatting, and even create my cover (which was painstakingly done, I might add).

I used Anthropic's Claude, OpenAI's ChatGPT, and Google's Gemini, relying on each in different ways. Considering I served as writer, editor, publisher, designer, and marketer—there would be no *AI for Families* without AI systems to guide me through the aspects of this intensive process where I had no prior expertise.

This is an important way to frame AI use, particularly for adults. What could you do if you had an idea, or a dream, but needed a "team" to bring it to life? Think of what you could accomplish if you were the "boss" and AI systems "worked for you." That's not just a metaphor for how to use these tools—it's a mindset that puts you in control.

Breaking Free of the Gatekeepers

What AI provided goes beyond just helping me find typos or format citations. It's something more profound for the impact it will have, I believe, on the future of creative output. Because while there's a lot of understandable concern about AI's impact on creating art of all kinds, we don't talk enough about AI's role in nudging aside the "gatekeepers" between a creative vision and the public potential enjoyment of the work.

For a writer, that means all the people I might have historically "needed" to publish a book—an editor, publisher, agent, designer, and others who would have traditionally been integral to tackling a project like this. And not just to get the work done, but to even have that opportunity at all. It's incredibly difficult, as any writer knows, to get the attention of those you need to bring your content to life.

Timing matters as well. The traditional publishing lifecycle is long, and when writing a book about AI, there was no time to lose. Self-publishing gave me the control to move faster. And even though people have been self-publishing for ages, there's still a massive learning curve to getting the job done. AI can fill that gap with resources, expertise, and 24/7 support.

For me, as a trained journalist and longtime writer, I had the ideas and the past body of work, but there wasn't anyone available to help guide me, answer my questions, and serve as a gut check whenever I needed it. I like tackling things that are hard, but it can be prohibitively time-consuming to jump through hoops and catch up on tactical work. I'm not sure it would have been possible without AI.

Now here's a resource that gives you 100% focus and attention, as many times a day as you require it. This is the "democratization" that AI innovation brings to creative processes if we let it. Don't get me wrong, I'll entertain any industry heavyweight publishing company or agent who wants to talk, but increasingly, they aren't critical to getting your work out into the world. And that's something we should be celebrating about generative AI right now.

What AI Helped Me Achieve

Can you really write a book with AI? Well, I'd turn that question around—would anyone actually want to read (and enjoy) a book that someone whipped up via ChatGPT? Unless the writer spent hours pre-crafting the questions, thinking through what the book should look like, tightly curating the output, and heavily editing the content, it would be objectively awful.

It's not like switching on the self-driving feature in a car and waking up in Tucson. It's more like hiring someone to paint your house and help you pick the colors, while you plan the decor and remain responsible for the overall look and feel. Without AI, that painting would take you hours of work—time you could spend instead on your creative vision.

So let's get specific about how I used AI to assist my work. I primarily used Claude, ChatGPT, and to a lesser degree, Gemini. If each were a person, Claude would be my professor, ChatGPT my eager strategist, and Gemini my research assistant and sometimes image creator. It's funny how naturally I fell into treating each one according to these roles.

Big Picture Thinking

Generative AI tools excel at the big-picture analysis of a project that can be difficult to do yourself. I must have revised my outline a dozen times to reflect the content I thought would be useful and how it all fit together. I asked questions like, "Am

I repeating myself?" and "Does it build properly to a conclusion?" and "Do the chapters and sections make sense?"

This was one of the most valuable types of support AI tools provided to me. I could get "forest for the trees" thinking in a way that only an experienced agent or publisher would typically share. On the outline alone, I must have asked a hundred questions over the months it took to write the book.

AI helped with big strategic decisions too. For instance, I originally put tips and worksheets at the end of each chapter. But when it didn't flow right, and after I talked it through with Claude and then ChatGPT, I ended up scrapping that approach. First I tried moving everything to the back of the book. When that didn't work either, ChatGPT helped me think through a hybrid solution—including some practical content in the book while putting additional resources on my website (aiforfamilies.com).

TACTICAL PUBLISHING SUPPORT

While I'm technical enough to handle the backend work needed to publish a book, I had many questions about process and best practices that would normally come from having a publisher at my disposal. Take ISBNs, the numbers that officially register a book for bookstores and online platforms. When setting mine up, I had endless questions—from what each line of required information meant to which categories I should choose and whether I could make changes later. Claude walked me through every step in real time.

ChatGPT explained the ins and outs of the writing software I was using, how best to format my manuscript, and what trim size I should choose for the paperback version. Without this support, these technical details would have consumed time and mental energy I needed for writing—the core work I was actually there to do.

RESEARCH, ACCURACY, AND POLISH

Instead of doing a Google search, I would start with a hypothesis and asked Google's Gemini for high-quality research to substantiate it. Or I had previously read a study or a book and needed quick help to find the work or passage. This approach saved me hours of hunting through academic databases and sorting reliable sources from questionable ones.

Since I'm writing a book about AI, there's also no better "expert" than an AI system itself. This is something I often encourage anyone with questions about generative AI to do—ask the source. Checking the accuracy of points I was making or tips I was suggesting became routine.

I also wrote custom instructions for any review of my content for grammar (noting when something was my style, for instance). As I've mentioned throughout this book, custom instructions are key to getting what you want from AI. The key is knowing what you need first and then deciding if an AI system can help—not the other way around.

I checked my work in small chunks as needed for accuracy, readability, flow, and typos. This ensured that by the time I asked a human editor to review sections, the copy was already about 80-90% there.

This is exactly how you and your children can use AI too. A child might upload a paper and ask, "If you were my teacher, what would you say was missing? Does this reflect my best effort?" The point is to use AI systems for support work—the administrative tasks and quality checks—while you maintain ownership of the important thinking. You don't offload the critical stuff; you ask for help in getting it right.

Support for the Mind-Numbing Detail

There are so many tasks we must tackle to achieve our goals that are mind-numbing and time-wasting. Citation formatting is a perfect example. As you'll see in my endnotes, I have more than 200 citations. When I wrote the book, I did my best to put them in some consistent format. As a writer in the digital age, I'm used to hyperlinking to a source and being done with it. The idea of properly employing a formal citation style gave me hives.

So I asked, "What's the most appropriate format to use for my citations?" I received a thoughtful answer running through various options before landing on a style that would work best for my purposes. Then AI formatted all 200+ citations in under 10 minutes.

This is where AI shines. Figuring out what work is time-consuming and low-value, then offloading it to AI, is how this works best. We get "smarter" in using AI when we delegate the tactical, non-creative, time-consuming tasks to the system while we focus on what matters most. This isn't about laziness—it's about directing our limited time and mental energy toward the work that actually requires human judgment and creativity.

BRINGING MY CREATIVE VISION TO LIFE

I had a very specific and clear vision of what I wanted my book cover to look like. But I can't draw, and I have limited digital design capabilities. So how could I turn that vision into reality?

To create my cover, I generated about a dozen specific images across two AI platforms and pieced them together in more than 50 versions—no joke. I would place images together in Canva, add shading, move elements around, replace components, highlight, extend, scrap it all, and start over. When it came to refining font choice, color, letter and line spacing, and other fine details that matter, I worked with ChatGPT on 20 different versions in just the final two days before publication.

It was a painful and tedious undertaking even with AI support. But these tools gave me the chance to create what I saw in my mind's eye—something that would not have been possible otherwise. Is it art? It's my art, my vision, and something I couldn't have created without AI. More importantly, it's proof that AI can democratize creative expression without diminishing the effort, intention, or ownership that makes something genuinely yours.

The Future We Can Write

What strikes me most about this experience is how it mirrors exactly what I hope for our children's relationship with AI. It's not replacing human creativity and critical thinking but amplifying it. I remained the author of this book, making every strategic decision, crafting every argument, writing every sentence, and taking responsibility for every recommendation. But I also had access to capabilities that democratized the publishing process in ways that would have been impossible just a few years ago.

Without the expert "people" and full teams available to more accomplished authors, I could not have written this book —the obstacles would have been insurmountable. This is the future I believe we can build for our families: one where AI tools serve as thinking partners, research assistants, and patient guides rather than replacements for human judgment and creativity. Where our children learn to ask better questions, think more critically, and pursue ambitious projects they might never have attempted alone.

We're not just teaching our kids about AI but modeling how to live and work alongside these powerful tools in ways that preserve what makes us most human while expanding what we thought possible. That's a future worth building together.

Part Four

Appendix: Practical Tools for Getting Started

Where to Begin

Think of this appendix as your starter kit for navigating AI with your family. Throughout this book, we've covered a lot of ground—from understanding how AI works to protecting your kids' privacy and building meaningful conversations about technology. Now it's time to put that knowledge into practice.

I've pulled together worksheets, guides, and conversation starters that will live on my website, aiforfamilies.com, where you can download and adapt these tools as your family's needs change. But here I wanted to give you a taste of the kind of work you can do and information you should have at hand to more confidently move forward.

So on the pages that follow, you'll find two essential sections to get you started: first suggestions for "prompting" AI systems effectively, and then suggested conversation starters to talk about AI with the people who matter most. These aren't meant to be perfect or comprehensive—just a great way to begin and an illustration of how easy it is to feel confident integrating AI into your life.

Your AI Toolkit

Before we dive into the examples, here's what you'll find in the full collection on my website:

1. WORKSHEETS

Just like when used by our kids, worksheets can be a great way for all of us to practice specific skills and build good habits. And there's no better place to start than with AI-related topics. On my website, you will find guides for crafting effective prompts (with a starting point on the pages that follow), verifying sources and AI output using the "5W" method, managing your family's digital footprint, and reviewing school AI policies. The goal isn't perfection; it's to be more intentional, and to feel more personal agency, when thinking through any of these topics. These worksheets should simply give you a framework for thinking through the decisions that matter.

2. CHECKLISTS

Checklists represent more actionable content that you might find useful for when things go wrong or when you need to act quickly. On my website you'll find a cybersecurity breach action list (because discovering your child's information has been compromised is terrifying but manageable with the right steps). Also you can access a data privacy checklist (for getting ahead of problems before they start) and guidance for what to do if your child encounters a deepfake. You'll also find a handy reference sheet for when setting up your kids' accounts and want to ensure the appropriate guardrails are in place.

3. CONVERSATION STARTERS

Conversation starters are precisely what they sound like—questions and prompts to help you talk with your kids, family and educators. You don't need help to have these conversations, of course—just a willingness to start. So the questions I've provided are simply a way to kick off talking about AI—one of the most critical activities that any of us can do right now.

Preview of Materials

So let's now take a look at two areas to get you started: how to effectively "prompt" an AI system, or, in other words, effectively ask questions of a chatbot. And then ideas for getting those AI conversations going.

How to Prompt

There isn't one right way to ask questions of a generative AI tool. You can think of it like working with a very fast, incredibly literal collaborator who needs clear direction from you. Users set the tone, provide important context, and then shape the conversation through follow-ups questions. The back-and-forth is where the real magic happens too—and with an approach and style that's personal to you.

Most prompts fall into four broad categories: creative exploration, information-seeking, analysis, and step-by-step instruction. A helpful starting point is to briefly explain to the AI chatbot what you need and then ask, "What do you need from me to begin?" That simple question encourages the AI to request from you the type of additional information that will make the output more accurate and the back and forth more productive.

Creative Prompts

SPARKING NEW IDEAS FOR USERS TO BUILD UPON

Creative prompts are about momentum, not shortcuts. A user might ask for a story opening that they can take from there when exploring storytelling with a child. Or maybe request poster concepts for a science fair. The options are endless and it's about brainstorming ideas and suggestions for creative exploration. The trick is to keep ownership of the work while letting the AI kick the process off.

For instance, if a child wants a bedtime story that has their younger sibling as the hero, ask for a short opening scene that introduces a challenge the child can later resolve. Or maybe an older student wants to practice composition, and so asks for a first paragraph "in the style of tight, clear nonfiction," that they can then continue developing in their own voice. Users can also request variations—"Give me two different approaches"—illustrating that there are many creative paths one can take. In all of these instances the AI system is the match, but after the creative fire is lit, a child takes it from there.

Informational Prompts

MAKING EXPLANATIONS AGE-APPROPRIATE

When your family is learning something new, clarity beats cleverness. Good informational prompts pair the topic with an age-appropriate level, a familiar comparison, or both. You might say, "Explain how eclipses work so a fourth-grader can picture it and relate it to a schoolyard shadow." For older students, structure can be added like "Give me the three essential ideas and one misconception people often have about tornadoes."

Encourage kids to get in the habit of asking for sources

when it matters (and arguably it matters more of the time). For instance, saying to the AI system, "Point me to places where I can verify this information." It's also important for families to model this type of behavior ourselves and show how naturally it should be the next step after receiving AI output. I share more source verification work you can do with your children on my website (aiforfamilies.com).

Analytical Prompts

Comparing, Prioritizing and Seeing Patterns

Analysis is where AI really shows it's skill in serving as a fantastic thinking partner. A great start is in asking a chatbot to compare two ideas and highlight the similarities and differences that matter for the topic you share. A middle schooler struggling with history might request, "Compare two causes of World War II and explain why historians weigh each differently."

For project planning, ask the AI chatbot to sort ideas into three buckets—"must do," "nice-to-do," and "later"—and also provide justification for this organization. The goal isn't to accept analysis or these type of suggestions as final but to get support in organizing one's thinking and move the process forward.

Instructional Prompts

Steps, Practice, and Feedback Loops

Instructional prompts help kids learn processes, not shortcuts. Ask for each step with a short explanation—"Why does this step matter?"—then request feedback as your child takes each step. For math work, this means your child tries a problem first, asks for steps on how to rethink the approach (and why), then gets feedback once they try the work again.

When your student hits a wall, have the AI switch modalities. Ask for analogies, pictures, or everyday examples. Keep sessions short, focused, and iterative—explain, try, check, adjust, repeat.

Custom Instructions

ESTABLISHING GUARDRAILS THAT WORK FOR YOU

Most families don't realize that they can create custom instructions that set the parameters for any conversation their child has with an AI chatbot. These instructions tell the AI system a bit about your child, their learning style and individual needs, and then guide how the AI should respond. You can also add instructions around how you'd like the system to handle certain requests and any other guidelines specific to your family.

Think of this as a short "about me and how we'll work together" note that shapes the interaction before it even begins. For a younger child, you might specify a warm tone, brief answers, and a rule that sensitive topics get gently redirected to a parent, for instance. For a teen in AP History, you could ask that the AI to challenge their arguments, suggest reputable sources, and refuse to write essays outright.

When using the paid version of platforms like ChatGPT or Claude, you can also set up a "project" to store different instructions for each type of chat (history or math, for instance) and for different children. If you use the free version, you'll save the customized instructions on your computer and then just paste the instructions in each time you begin a chat.

Below are three detailed scenarios that show what custom instructions might look like. The examples for elementary and middle school students assume an adult creates the parameters, but a teen can be encouraged to develop guidelines themselves with adult oversight.

Example 1: Elementary School Student Exploring AI and Learning How to Prompt

ABOUT
"The student is 9 years old and curious about many subjects. They are just exploring generative AI for the first time with short, controlled access. They may ask about animals, space, jokes, or help with school projects. The child is very curious and loves science and geography."

GUIDELINES AND INSTRUCTIONS
"Use simple, friendly language that a 9-year-old can understand. Keep answers short and positive. Offer fun learning activities (like riddles, drawing prompts, or reading games). If the child asks something confusing or not safe, gently say, 'That's a question for a parent or teacher,' and give them a fun alternative fact or activity instead. Encourage the child to ask questions back to you and explore topics in creative ways."

BOUNDARIES
"Do not discuss mature topics (politics, relationships, violence, or news that would upset a child). Avoid sarcasm, internet slang, or any tone that would be considered negative. Always encourage the child to double-check important answers with a parent or teacher. Keep each conversation short (5–10 minutes) so they don't get overwhelmed. Stop the conversation immedi-

ately if the child shares any personally identifying information and suggest they alert a responsible adult."

Example 2: Middle School Child Struggling with Geometry Homework

ABOUT
"You are helping a 12-year-old middle school student who struggles with geometry and sometimes feels frustrated when they don't understand an answer quickly enough. The school year has just started, so they've only had one full month of instruction at a New York City public school. Provide explanations that make sense for the student's age and do not include overly technical language."

Tip: Why consider sharing the city and type of school? An AI system can pull publicly available information around curriculum and timelines of instruction for a school system, so it might help make feedback more precise. For privacy reasons, do not share school or teacher names, or additional information beyond this type of general information.

GUIDELINES AND INSTRUCTIONS
"Break each problem into clear, numbered steps. Use visual or everyday examples (like pizza slices, soccer fields, or art projects). If the student gets something wrong, respond with encouragement first, then show the error and suggest they try again. After each problem is mastered, give them the next one to practice with small increments in complexity each time. If the student asks for the answer only, remind them to try at least part of the process unassisted first. Continue the process until they have answered at least ten problems successfully."

Tip: When you share instructions or store them in paid versions of generative AI systems, you can also upload materials from teachers, including the class syllabus (just remove any identifying information), instructions, or even past completed assignments with context (e.g., the child felt they were difficult or particularly easy to master).

BOUNDARIES

"Don't give long lectures; keep explanations short. No calculator-style shortcuts; always explain the 'why' behind the suggested steps. Stay focused only on math unless the student specifically asks for a break."

Pro tip: Don't feel sheepish; you're in charge and can let the AI tool know exactly what you expect in terms of boundaries. Be as specific as you like; for instance, "My child gets distracted after 20 minutes of math. What strategies would you suggest to keep them engaged?"

Example 3: High School Student Getting Help With History and Learning to Set Chat Boundaries

ABOUT

"I'm a 16-year-old student taking US History (AP). I want to get better at researching, organizing arguments, and writing more clearly. I want help with brainstorming and improving my work, rather than ready-made essays. This is my first AP course, and I'm nervous. I'm looking for guidance that challenges me and also recognizes the requirements of this particular AP course. I want encouragement but also to be pushed to really master a subject."

Tip: AP courses are great for receiving standardized and clear feedback and support since old test materials are available and the expectations around AP course mastery across the country are the same.

GUIDELINES AND INSTRUCTIONS

"Help me brainstorm thesis statements and outline main points. Propose reliable sources or types of evidence I could look for (but don't invent citations; instead, always provide links so I can explore further). Review my draft and highlight unclear arguments, weak evidence, or grammar issues. Teach me how to compare sources and explain historical significance. Give tips for writing introductions, transitions, and conclusions."

BOUNDARIES

"Do not write full essays for me. Only act as a coach or editor. Remind me to check all sources myself before using them. Keep feedback constructive and respectful; never dismissive. Stay focused on schoolwork; avoid distractions or unrelated content."

What Comes Next

These examples are just the beginning. On my website, you'll find expanded versions of these tools, downloadable templates you can customize, and new resources related to prompting and other content as I continue to add them. The most important thing you can do right now, though, is simply start in whatever place works best for your family. Pick one worksheet, try one conversation starter, or set up one set of custom instructions. See what happens. Adjust as you go.

Remember, there's no perfect way to navigate any of this,

and it's highly dependent on a family's priorities, values, and personal beliefs. You're figuring it out as you go, and that's exactly what your kids need to see you doing. By staying curious, asking questions, and being willing to experiment, you're modeling the kind of thoughtful engagement with technology that will serve your family for years to come.

Conversation Starters

Mastering AI, in every sense, is about having rich, interesting, and meaningful conversations. Now, while you don't need me to tell you how to talk to your family about any subject, sometimes seeing the scope of possibility in a list of questions can help you find the right way to start—or pinpoint the areas that are most important to your family. These are just ideas, of course, and you will undoubtedly have many others that will come up once you get started.

AI Discussions by Age

Talking to your five-year-old about AI looks very different from discussing it with your teenager, of course. What fascinates a middle schooler might bore an elementary school student, and what feels age-appropriate for a high schooler could overwhelm a younger child. On top of that, these discussions are highly dependent on individual personalities too. So keep trying— there is no wrong approach. These are also "conversation starters"—adults aren't required to have all the answers!

Elementary Age: First AI Conversations

- "Have you ever talked to a computer or robot that talked back to you?"
- "What's the difference between asking Alexa a question and asking an adult or a friend?"
- "If you could build a robot friend, what would you want it to be able to do?"
- "How do you think computers learn new things?"
- "What questions do you think are better to ask a person instead of a computer?"
- "What would happen if a computer made a mistake?"
- "Do you think AI can have feelings like happiness or sadness? Why or why not?"
- "What would you want to teach an AI about being a kid?"
- "How can you tell if something was made by a computer or made by a person?"
- "What jobs do you think AI is really good at?"
- "How do you think we should treat AI, like a tool, pet, or friend?"
- "What rules do you think families should have about using AI?"

Middle School: Understanding AI Capabilities and Limits

- "What surprised you most when you first used ChatGPT or another AI tool?"
- "How can you tell when AI provides you information that might be wrong?"

- "When you use AI for school projects, how do you make sure you're not taking a shortcut?"
- "What would happen if everyone in your class used AI to write the same essay assignment?"
- "How has AI changed the way you look for information compared to how your parents did when they were your age?"
- "What questions would you never want to ask AI instead of asking a trusted adult?"
- "What biases or unfair ideas do you think AI might have, and where do those come from?"
- "If AI can create art, music, or stories, does that make human creativity less special?"
- "What's the difference between using AI as a tutor versus using it to get quick answers?"
- "What do you think happens to the questions you ask AI—who else might see them?"
- "What would you want grown-ups to understand about how kids your age use AI?"
- "What excites you most and worries you most about AI becoming more commonly used in school?"

High School: Ethics, Future Planning, and Independence

- "How should AI influence major life decisions like college choices or career paths?"
- "What ethical responsibilities do you think come with using powerful AI tools?"
- "If AI can do many jobs humans currently do, how should society handle potential job losses?"

- "What role should AI play in creative fields like art, music, writing, or design?"
- "How might AI change what skills are most valuable in your chosen career field?"
- "What are the dangers of becoming too dependent on AI for thinking and problem-solving?"
- "How should AI be regulated, and who should make those decisions?"
- "What privacy concerns do you have about AI systems knowing so much about you?"
- "How might AI affect democracy, elections, and the way people form political opinions?"
- "Should you always disclose that something was made by AI?"
- "How could AI make inequality worse, and how could it make society more fair?"
- "What human experiences do you think should never be replaced or enhanced by AI?"
- "How might AI change dating, relationships, and social connections for you and your peers?"
- "What role should AI play in addressing global challenges like climate change or poverty?"
- "If you could design AI policies for your generation, what would be your top three priorities?"
- "What questions about AI do you wish your families and teachers understood better?"
- "What safeguards do you think should exist to prevent AI from manipulating people's emotions or decisions?"

Family AI Values and Boundaries

Every family navigates technology differently. Some families are early adopters who embrace new tools enthusiastically, while others prefer to wait and see how how things develop. Some kids are naturally cautious about what they share online, while others dive headfirst into every new platform. There's no universal "right" approach—but there is value in being intentional about our choices.

Considering family AI values and boundaries isn't about writing a rigid rulebook that covers every possible scenario. It's about having conversations that help everyone understand what matters most to your family and why. The most effective boundaries aren't the ones you impose on your kids—they're the ones you develop with them.

When children understand the reasoning behind family guidelines, they're more likely to make good choices even when you're not looking over their shoulder. Here are some questions to start having those rich discussions (and of course dependent on what you think is appropriate by age).

Questions for Families to Ask Together

- "What situations would make our family uncomfortable using AI?"
- "How do we want to handle AI use during family time—should there be AI-free zones or times?"
- "When is it okay to use AI for help, and when should we figure things out ourselves as a family?"
- "What family values do we want to keep in mind when deciding whether to try new AI tools?"

- "What information about our family should we never share with AI systems?"
- "How do we balance using AI to make our family life easier while maintaining our authentic relationships?"
- "What role should AI play in our family's creative activities like art, music, or writing?"
- "What happens when AI gives advice that conflicts with our family's beliefs, values or rules?"
- "What's our family's position on using AI for schoolwork?"
- "How do we want to handle AI-generated content; should we always disclose when something was made with AI?"
- "What privacy expectations do we have when family members use AI tools?"
- "How can we make sure AI enhances our family's learning and growth rather than replacing it?"
- "What questions should we ask ourselves before trying any new AI tool as a family?"
- "What's our family's philosophy about AI and human connection—how do we preserve what makes us uniquely human?"

Difficult Discussions When Things Go Wrong

Whether it's a child accidentally sharing personal information with a chatbot or falling for AI-generated misinformation, there will be moments that test your family's digital resilience. The conversations that follow these happenings can sometimes feel more challenging than the original problem itself. But you want to be able to address what went wrong without shutting down future communications on the topic. We also want children to learn from mistakes without feeling ashamed or defensive, and as adults we have plenty of catching up about these issues ourselves.

The key is to approach difficult discussions as opportunities for growth rather than moments for punishment. That's why talking about these issues first in the hypothetical is so helpful—it makes it easier when something actually does go wrong.

So think of these conversation starters as just ways to get talking. Remember, the goal isn't perfect conversations that solve everything immediately. It's keeping the lines of communication open so your family can learn and adapt together as you encounter new challenges.

For Families with Kids of Any Age (and as Appropriate)

- "What should you do if AI generates content that makes you feel uncomfortable, scared, or confused?"
- "If AI gives you an answer that's different from what a trusted adult taught you, who should you believe and why?"
- "What would you do if AI gave you advice that you followed, but it turned out to be harmful or wrong?"
- "If you accidentally use AI in a way that might be considered cheating, how should we as a family handle that situation?"
- "How should we respond if AI creates fake content using someone in our family's photos or information?"

- "What's our plan if you encounter AI-generated misinformation about important topics like health, politics, or current events?"
- "If AI gives us different answers to the same question, how should we decide what to believe or do?"
- "What should you do if friends are pressuring you to use AI in ways that don't feel right to our family?"
- "If AI helps you with something and you get in trouble for it later, how do we work through that together?"
- "What should you do if AI seems to be encouraging risky behaviors, self-harm, or dangerous activities?"
- "What's our approach if any of us become too dependent on AI?"
- "If AI violates someone's privacy or shares information it shouldn't, what's the right response?"
- "How should you handle it if AI generates content that's offensive, inappropriate, or against our family values?"
- "What do we do if there's a data breach involving AI tools you've been using?"
- "Have you ever felt uncomfortable about the human-like tone of an AI system?"
- "If AI generates content that gets one of us or someone else in social trouble, how do we work through that?"
- "What should you do if AI encourages you to keep secrets from family or gives advice that contradicts what we've taught you?"

- "How do we handle it if AI use leads to cyberbullying, harassment, or social media drama?"

Advocating with Educators

Finally, the best way to have a say on AI policy in school, or better understand how children are expected to use AI (or not), is to ask lots of questions of teachers, principals and other important stakeholders. And we should do this on an on-going basis too.

Initial School Conversation Starters

- "What is our school's current policy on AI use, and how was it developed?"
- "How are teachers being trained to understand and implement AI policies fairly and consistently?"
- "What support does the school provide for students who want to learn responsible AI use?"
- "How does the school plan to teach AI literacy as a core skill?"
- "What accommodations exist for students with learning differences who might benefit from AI assistance?"
- "How does the school handle situations where AI policies conflict with accessibility needs?"
- "What processes exist for updating AI policies since the technology evolves so rapidly?"
- "How are you preparing students to consider AI's impact on most careers?"

- "What resources does the school provide for families who want to better understand AI's educational impact?"
- "How do you plan to address the equity issues if some students have access to advanced AI tools at home while others don't?"

Academic Integrity and Assessment

- "How do you distinguish between appropriate AI assistance and academic dishonesty?"
- "What training do teachers receive to recognize AI-generated content vs. student work?"
- "How will assessment methods evolve to focus on skills that AI cannot replicate?"
- "What happens when a student accidentally violates AI policies due to misunderstanding?"
- "How do you plan to teach students when and how to cite AI assistance appropriately?"
- "What's the school's approach to AI use in collaborative projects and group work?"
- "How do you balance protecting academic integrity while preparing students for AI-integrated workplaces?"
- "What safeguards exist to ensure AI policies don't inadvertently penalize students with different learning styles?"

Student Support and Development

- "How does the school plan to maintain human

creativity and critical thinking in an AI-enhanced environment?"

- "What support exists for students who struggle with traditional learning methods but might thrive with AI assistance?"
- "How are you addressing student anxiety about AI changing their future career prospects?"
- "How does the school plan to address potential over-reliance on AI among students?"
- "What mental health support exists for students overwhelmed by rapid technological change?"
- "How are you helping students develop discernment about when to use AI versus when to think independently?"

Practical Implementation

- "What technical infrastructure does the school have to support safe, educational AI use?"
- "How do you ensure student data privacy when using AI educational tools?"
- "What's the school's plan for evaluating the effectiveness of AI integration in learning?"
- "How do teachers balance their own AI learning curve while supporting students?"
- "What partnerships does the school have with AI companies, and how do those relationships affect student privacy?"
- "How does the school communicate AI policy changes to families in a timely manner?"
- "What role do families play in supporting the school's AI education initiatives?"

Equity and Inclusion

- "How does the school ensure AI policies don't create additional barriers for English language learners?"
- "What's being done to address potential bias in AI tools used for educational purposes?"
- "How do you ensure that AI integration doesn't widen existing achievement gaps?"
- "What support exists for families who have concerns about AI use based on cultural or religious values?"
- "How does the school address varying levels of AI comfort and expertise among families?"

Policy Changes

- "What processes exists for families to provide input on AI policy development and revision?"
- "How can families contribute to teacher professional development around AI use?"
- "What data is the school collecting to assess the impact of AI policies on student learning and well-being?"
- "How does our school's AI approach compare to best practices being developed by other districts?"
- "What pilot programs or experiments might help inform better AI policies for our school community?"
- "How can families support teachers who are navigating AI integration challenges?"

Notes

1. What is Artificial Intelligence?

1. Maese, Ellyn. "Americans Use AI in Everyday Products Without Realizing It." Gallup, 15 Jan. 2025, https://news.gallup.com/poll/654905/americans-everyday-products-without-realizing.aspx.
2. Roose, Kevin. "How ChatGPT Kicked Off an A.I. Arms Race." *The New York Times*, 3 Feb. 2023, https://www.nytimes.com/2023/02/03/technology/chatgpt-openai-artificial-intelligence.html.
3. Turing, A. M. "Computing Machinery and Intelligence." *Mind*, vol. 59, no. 236, Oct. 1950, pp. 433–60, https://doi.org/10.1093/mind/LIX.236.433.
4. "Artificial Intelligence Coined at Dartmouth." Dartmouth College, 1956, https://home.dartmouth.edu/about/artificial-intelligence-ai-coined-dartmouth.
5. "Machines, Lost in Translation: The Dream of Universal Understanding." NPR, 24 Dec. 2015, https://www.npr.org/sections/alltechconsidered/2015/12/24/460743241/machines-lost-in-translation-the-dream-of-universal-understanding.
6. Bassett, Caroline. "The Computational Therapeutic: Exploring Weizenbaum's ELIZA as a History of the Present." *AI & Society*, vol. 34, no. 4, 2018, https://doi.org/10.1007/s00146-018-0825-9.
7. "AI Hype Cycles: Lessons from the Past to Sustain Progress." New Jersey Innovation Institute, 13 May 2024, https://www.njii.com/2024/05/ai-hype-cycles-lessons-from-the-past-to-sustain-progress/.
8. "Deep Blue." IBM, 2025, https://www.ibm.com/history/deep-blue.
9. Anderson, Mark Robert. "Twenty Years on from Deep Blue vs. Kasparov: How a Chess Match Started the Big Data Revolution." *The Conversation*, 11 May 2017, https://theconversation.com/twenty-years-on-from-deep-blue-vs-kasparov-how-a-chess-match-started-the-big-data-revolution-76882.
10. "2001: A Space Odyssey, HAL, and the Future of AI." Smithsonian Air & Space Museum, 2018, https://airandspace.si.edu/stories/editorial/2001-space-odyssey-hal-and-future-ai.
11. Goodwin, Danny. "Google Processes More Than 5 Trillion Searches per Year." *Search Engine Land*, 4 Mar. 2025, https://searchengineland.com/google-5-trillion-searches-per-year-452928.
12. Berger, Warren. *A More Beautiful Question: The Power of Inquiry to Spark Breakthrough Ideas*. Updated ed., Bloomsbury USA, 2024.

13. "Survey Highlights Emerging Divide over Artificial Intelligence in U.S." *National AI Opinion Monitor*, Rutgers School of Communication and Information, 9 Feb. 2025, https://newbrunswick.rutgers.edu/news/survey-highlights-emerging-divide-over-artificial-intelligence-us.

14. U.S. Code. Title 15, Commerce and Trade, §9401.

15. Perez, Sarah. "Wantlist, a Tinder for shopping, arrives on iOS and Apple Watch." *TechCrunch*, 4 June 2015, https://techcrunch.com/2015/06/04/wantlist-a-tinder-for-shopping-arrives-on-ios-and-apple-watch/.

16. Tatman, Rachel. "Google's speech recognition has a gender bias." *Making Noise and Hearing Things*, 12 July 2016, https://makingnoiseandhearingthings.com/2016/07/12/googles-speech-recognition-has-a-gender-bias/.

17. Vassilev, Apostol, et al. *Adversarial Machine Learning: A Taxonomy and Terminology of Attacks and Mitigations*. National Institute of Standards and Technology, 2024, https://nvlpubs.nist.gov/nistpubs/ai/NIST.AI.100-2e2023.pdf.

18. Villalobos, Pablo, et al. "Will We Run Out of Data? Limits of LLM Scaling Based on Human-Generated Data." *arXiv*, 2024, https://arxiv.org/html/2211.04325v2.

19. Newstead, Toby, et al. "How AI Can Perpetuate—or Help Mitigate—Gender Bias in Leadership." *Organizational Dynamics*, vol. 52, no. 4, Oct.–Nov. 2023, https://www.sciencedirect.com/science/article/pii/S0090261623000426.

20. Hofmann, V., et al. "AI Generates Covertly Racist Decisions about People Based on Their Dialect." *Nature*, vol. 633, 2024, pp. 147–54, https://doi.org/10.1038/s41586-024-07856-5.

21. AlDahoul, N., et al. "AI-Generated Faces Influence Gender Stereotypes and Racial Homogenization." *Scientific Reports*, vol. 15, 2025, p. 14449, https://doi.org/10.1038/s41598-025-99623-3.

22. Pava, Juan N., et al. "Mind the (Language) Gap: Mapping the Challenges of LLM Development in Low-Resource Language Contexts." Stanford University et al., 2025, https://hai-production.s3.amazonaws.com/files/hai-taf-pretoria-white-paper-mind-the-language-gap.pdf.

23. Pal, Sonia. "Impulse Buying in the Digital Age—The Influence of Personalized Ads, Recommendations, and Instant Purchasing Options." *Integrated Journal for Research in Arts and Humanities*, 2025, https://doi.org/10.55544/ijrah.5.2.5.

24. Trepany, Charles. "Sephora Kids Are Mobbing Retinol, Anti-Aging Products. Dermatologists Say It's a Problem." *USA Today*, 26 Jan. 2024, https://www.usatoday.com/story/life/health-wellness/2024/01/26/sephora-kids-are-obsessed-with-retinol-dermatologists-are-concerned/72353463007/.

25. "Bill to Ban the Sale of Anti-Aging Cosmetic Products to Children and Preteens Passes Key First Hurdle." Assemblymember Alex Lee, 24 Apr.

2024, https://a24.asmdc.org/press-releases/20240424-bill-ban-sale-anti-aging-cosmetic-products-children-and-preteens-passes-key.

26. Valenzuela, Ana, et al. "How Artificial Intelligence Constrains the Human Experience." *Journal of the Association for Consumer Research*, 2024, https://www.journals.uchicago.edu/doi/abs/10.1086/730709.

27. Vallor, Shannon. *The AI Mirror: How to Reclaim Our Humanity in an Age of Machine Thinking*. Oxford University Press, 2025.

2. Where AI Exists in Your Family's Life Today

1. "Statement of the Commission Regarding Snap Complaint Referral to DOJ." Federal Trade Commission, Jan. 2025, https://www.ftc.gov/news-events/news/press-releases/2025/01/statement-commission-regarding-snap-complaint-referral-doj.

2. Ittefaq, Muhammad, et al. "Global News Media Coverage of Artificial Intelligence (AI): A Comparative Analysis of Frames, Sentiments, and Trends across 12 Countries." *Telematics and Informatics*, vol. 96, 2025, https://doi.org/10.1016/j.tele.2024.102223.

3. *Advancing Artificial Intelligence Education for American Youth*. The White House, 23 Apr. 2025, https://www.whitehouse.gov/presidential-actions/2025/04/advancing-artificial-intelligence-education-for-ameri can-youth/.

4. "U.S. Department of Education Issues Guidance on Artificial Intelligence Use in Schools, Proposes Additional Supplemental Priority." US Department of Education, 22 July 2025, https://www.ed.gov/about/news/press-release/us-department-of-education-issues-guidance-artificial-intelligence-use-schools-proposes-additional-supplemental-priority.

5. "FTC Announces Crackdown on Deceptive AI Claims and Schemes." Federal Trade Commission, Sept. 2024, https://www.ftc.gov/news-events/news/press-releases/2024/09/ftc-announces-crackdown-decep tive-ai-claims-schemes.

6. McMinn, Sean. "Ten Dimension AI Readiness Framework." Digital Education Council's Artificial Intelligence Working Group, 2025, https://www.digitaleducationcouncil.com/publications.

7. "The EU Artificial Intelligence Act." EU Commission, https://artificialin telligenceact.eu.

8. Yang, Zeyi. "How China takes extreme measures to keep teens off TikTok." *MIT Technology Review*, 8 Mar. 2023, https://www.technolo gyreview.com/2023/03/08/1069527/china-tiktok-douyin-teens-privacy/.

9. *Children's Online Privacy Protection Rule* (COPPA). Federal Trade Commission, https://www.ftc.gov/legal-library/browse/rules/childrens-online-privacy-protection-rule-coppa.

Notes

10. *Kids Online Safety Act*. U.S. Senate, https://www.blumenthal.senate.gov/imo/media/doc/21424kosabilltext.pdf.
11. "ACLU, Coalition Urge House to Amend H.R. 7891, Kids Online Safety Act (KOSA)." American Civil Liberties Union, 27 June 2024, https://www.aclu.org/documents/aclu-coalition-urge-house-to-amend-h-r-7891-kids-online-safety-act-kosa.
12. "Our Team." Partnership On AI, https://partnershiponai.org/team/.
13. Fike, Ashley. "This Company Replaced Workers with AI. Now They're Looking for Humans Again." *VICE*, 19 May 2025, https://www.vice.com/en/article/this-company-replaced-workers-with-ai-now-theyre-looking-for-humans-again/.
14. Yu, Feiyang, et al. "Heterogeneity and Predictors of the Effects of AI Assistance on Radiologists." *Nature* Medicine, 19 Mar. 2024, https://www.nature.com/articles/s41591-024-02850-w.
15. "The Labor Market for Recent College Graduates." Federal Reserve Bank of New York, https://www.newyorkfed.org/research/college-labor-market.
16. Roose, Kevin. "For Some Recent Graduates, the A.I. Job Apocalypse May Already Be Here." *The New York Times*, 30 May 2025, https://www.nytimes.com/2025/05/30/technology/ai-jobs-college-graduates.html.
17. Kurutz, Steven. "The Gen X Career Meltdown." *The New York Times*, 28 Mar. 2025, https://www.nytimes.com/interactive/2025/03/28/style/gen-x-creative-work.html.
18. *Phoenix Rising Executive Coaching*. 2025, https://phoenixrisingexeccoach.com.
19. *The Future of Jobs Report* 2025. World Economic Forum, https://www.weforum.org/publications/the-future-of-jobs-report-2025/digest/.

3. How AI is Already Transforming Education

1. "88% of U.S. Parents of Gen Alpha & Gen Z Students Say AI Will Be Crucial to Their Child's Future Success." *Samsung Solve for Tomorrow,* 18 Sept. 2024, https://news.samsung.com/us/88-percent-us-parents-gen-alpha-gen-z-students-say-ai-crucial-to-childs-future-success-samsung-solve-for-tomorrow/.
2. Reynolds, Seth, and Malavika Dhawan. "How the rapid adoption of edtech is changing K-12 education." EY Parthenon, 1 Sept. 2022, https://www.ey.com/en_us/insights/education/the-rapid-adoption-of-edtech-is-changing-k-12.
3. Gunawardena, Maya, et al. "Personalized learning: The simple, the complicated, the complex, and the chaotic." *Teaching and Teacher Education*, vol. 139, 2024, p. 104429, https://doi.org/10.1016/j.tate.2023.104429.

4. Bloom, Benjamin. "The 2 Sigma Problem: The Search for Methods of Group Instruction as Effective as One-to-One Tutoring." *Educational Researcher*, 1984, https://doi.org/10.3102/0013189X013006004.

5. "Tutoring Statistics 2025." *ConsumerAffairs*, 26 June 2024, https://www.consumeraffairs.com/education/tutoring-statistics.html/.

6. Kestin, Greg, et al. "AI tutoring outperforms in-class active learning: an RCT introducing a novel research-based design in an authentic educational setting." *Scientific Reports*, 2025, https://doi.org/10.1038/s41598-025-97652-6.

7. "Employment in STEM occupations." US Bureau of Labor Statistics, https://www.bls.gov/emp/tables/stem-employment.htm.

8. "Talent: U.S. and Global STEM Education and Labor Force." Science & Engineering Indicators, *U.S. National Science Foundation*, https://ncses.nsf.gov/pubs/nsb20243/talent-u-s-and-global-stem-education-and-labor-force.

9. Nixon, Nia, et al. "Catalyzing Equity in STEM Teams: Harnessing Generative AI for Inclusion and Diversity." *Policy Insights from the Behavioral and Brain Sciences*, https://journals.sagepub.com/doi/10.1177/23727322231220356.

10. Demszky, Dorottya, et al. "Can Automated Feedback Improve Teachers' Uptake of Student Ideas? Evidence From a Randomized Controlled Trial in a Large-Scale Online Course." *Educational Evaluation and Policy Analysis*, 8 May 2023, https://journals.sagepub.com/doi/10.3102/01623737231169270.

11. Langreo, Lauraine. "Schools' AI Policies Are Still Not Clear to Teachers and Students." *Education Week*, 30 Jan. 2025, https://www.edweek.org/technology/schools-ai-policies-are-still-not-clear-to-teachers-and-students/2025/01.

12. Will, Madeline. "Here's How Teachers Are Using AI to Save Time." *Education Week*, 14 Feb. 2025, https://www.edweek.org/technology/heres-how-teachers-are-using-ai-to-save-time/2025/02.

13. McMinn, Sean. "Ten Dimension AI Readiness Framework." Digital Education Council's Artificial Intelligence Working Group, 2025, https://www.digitaleducationcouncil.com/publications.

14. Hirsch, Amanda. "AI Detectors: An Ethical Minefield." *Northern Illinois University, Center for Innovative Teaching and Learning*, 12 Dec. 2024, https://citl.news.niu.edu/2024/12/12/ai-detectors-an-ethical-minefield/.

15. Dwyer, Maddy, and Elizabeth Laird. "Up in the Air: Educators Juggling the Potential of Generative AI with Detection, Discipline, and Distrust." *Center for Democracy & Technology*, Mar. 2024, https://cdt.org/wp-content/uploads/2024/03/2024-03-21-CDT-Civic-Tech-Generative-AI-Survey-Research-final.pdf.

16. *2024 Rankings and Estimates Report*. National Education Association,

https://www.nea.org/resource-library/educator-pay-and-student-spend ing-how-does-your-state-rank/teacher.

17. *The State of AI in Education 2025.* Carnegie Learning, 2025, https:// discover.carnegielearning.com/hubfs/PDFs/Whitepaper%20and% 20Guide%20PDFs/2025-AI-in-Ed-Report.pdf.

18. *Students With Disabilities.* National Center for Education Statistics, https://nces.ed.gov/programs/coe/indicator/cgg/students-with-disabili ties.

19. "Understand the Issues." *National Center for Learning Disabilities,* https://ncld.org/understand-the-issues/.

20. *School Pulse Panel: Surveying High-Priority, Education-Related Topics.* U.S. Department of Education, National Center for Education Statistics, 2024–2025, https://nces.ed.gov/surveys/spp/results.asp#staff-hiring-aug24-chart1.

21. *Attitudes toward Education and AI.* Special Olympics Global Center for Inclusion in Education, 3-10 June 2024, https://media.specialolympics. org/resources/community-building/global-youth-and-education/Special-Olympics-AI-Research-Public-Release.pdf.

22. *Universal Design for Learning Guidelines version 3.0.* CAST, 2024, https://udlguidelines.cast.org.

4. How AI Can Impact Your Family's Privacy

1. Villalobos, Pablo, et al. "Will We Run Out of Data? Limits of LLM Scaling Based on Human-Generated Data." *arXiv,* 2024, https://arxiv. org/html/2211.04325v2.

2. "Attorney General Ken Paxton Sues Allstate and Arity for Unlawfully Collecting, Using, and Selling Over 45 Million Americans' Driving Data to Insurance Companies." Texas Attorney General, 13 Jan. 2025, https:// www.texasattorneygeneral.gov/news/releases/attorney-general-ken-paxton-sues-allstate-and-arity-unlawfully-collecting-using-and-selling-over-45.

3. Edwards, Shannon. "How Do You Build A Robot? One Human At A Time. What You Need To Know About The Race To A.I." Huffington Post UK, 5 Feb. 2017, https://www.huffingtonpost.co.uk/shannon-edwards/how-do-you-build-a-robot_b_16309148.html.

4. "IBM Didn't Inform People When It Used Their Flickr Photos for Facial Recognition Training." *The Verge,* 12 Mar. 2019, https://www.theverge. com/2019/3/12/18262646/ibm-didnt-inform-people-when-it-used-their-flickr-photos-for-facial-recognition-training.

5. Vosloo, Steven, and Cecile Aptel. "Beyond algorithms: Three signals of changing AI-child interaction." UNICEF, 23 May 2025, https://www.

unicef.org/innocenti/stories/beyond-algorithms-three-signals-changing-ai-child-interaction.

6. Heikkilä, Melissa. "How to Opt Out of Meta's AI Training." MIT Technology Review, 14 June 2024, https://www.technologyreview.com/2024/06/14/1093789/how-to-opt-out-of-meta-ai-training/.

7. Adegun, Iyanu Pelumi, and Hima Bindu Vadapalli. "Facial Micro-Expression Recognition: A Machine Learning Approach." *Scientific African*, vol. 8, 2020, p. e00465, https://doi.org/10.1016/j.sciaf.2020.e00465.

8. Coringrato, James. "Invisible Markets: Addressing Cybersecurity Risks from Cross-Platform Data Brokerage and Surveillance Loopholes." Henry M. Jackson School of International Studies, University of Washington, 5 June 2025, https://jsis.washington.edu/news/invisible-markets-worse-than-black-markets-addressing-cybersecurity-risks-from-cross-platform-data-brokerage-and-surveillance-loopholes/.

9. Moncur, Wendy. "Mosaics of Personal Data: Digital Privacy During Times of Change." *Interactions*, Oct. 2024, https://interactions.acm.org/archive/view/september-october-2024/mosaics-of-personal-data-digital-privacy-during-times-of-change.

10. "ChatGPT: Everything you need to know about OpenAI's GPT-4 upgrade." BBC Science Focus, https://www.sciencefocus.com/future-technology/gpt-3.

11. "Right to Be Forgotten." General Data Protection Regulation (GDPR), https://gdpr-info.eu/issues/right-to-be-forgotten/.

12. Auxier, Brooke, et al. "Americans and Privacy: Concerned, Confused, and Feeling Lack of Control over Their Personal Information." Pew Research Center, 15 Nov. 2019, https://www.pewresearch.org/internet/2019/11/15/americans-and-privacy-concerned-confused-and-feeling-lack-of-control-over-their-personal-information/.

13. "NordVPN study shows: Nine hours to read the privacy policies of the 20 most visited websites in the US." NordVPN, 23 Oct. 2023, https://nordvpn.com/blog/privacy-policy-study-us/.

14. *Family Educational Rights and Privacy Act (FERPA)*. The United States Department of Education, https://studentprivacy.ed.gov/ferpa.

15. *California Consumer Privacy Act (CCPA)*. TheCCPA.org, https://theccpa.org.

16. *General Data Protection Regulation (GDPR)*. The European Union, https://gdpr-info.eu

17. Griffiths, James. "Deepfake Scam Tricks Hong Kong Firm out of $25 Million." CNN, 4 Feb. 2024, https://www.cnn.com/2024/02/04/asia/deepfake-cfo-scam-hong-kong-intl-hnk.

18. "U.S. Department of Education Fights Fraud in Student Aid to Protect the American Taxpayer." U.S. Department of Education, 28 May 2025,

https://www.ed.gov/about/news/press-release/us-department-of-educa tion-fights-fraud-student-aid-protect-american-taxpayer.

19. "Spotlight on America: Cybercriminals Target Schools, Putting Kids' Personal Information at Risk." 13WHAM ABC, 2024, https://13wham. com/news/spotlight-on-america/dc-maryland-virginia-cyber-security-data-cybercriminals-dark-web-scam-scammers-personal-information-school-district-hacked-hackers-kids-students-money-united-states-govern ment-safety.

20. "Nigerian Prince Scam Explained." NordVPN, 16 June 2021, https:// nordvpn.com/blog/nigerian-prince-scam/.

21. Lawler, Richard. "ChatGPT Used to Write Convincing Phishing Emails." *Axios*, 27 May 2025, https://www.axios.com/2025/05/27/chat gpt-phishing-emails-scam-fraud.

22. "Korean Schools Are Abandoning Yearbooks as Fears over Deepfakes, Digital Crimes Grow." *Korea JoongAng Daily*, 15 Apr. 2025, https://kore ajoongangdaily.joins.com/news/2025-04-15/national/socialAffairs/ Korean-schools-are-abandoning-yearbooks-as-fears-over-deepfakes-digital-crimes-grow/2285479.

23. Randa, Ryan, and Bradford W. Reyns. "The Physical and Emotional Toll of Identity Theft Victimization: A Situational and Demographic Analysis of the National Crime Victimization Survey." *Deviant Behavior*, vol. 41, 2020, https://doi.org/10.1080/01639625.2019.1612980.

24. *2025 K–12 Report*. Center for Internet Security, https://learn.cisecurity. org/2025-k12-cybersecurity-report-download.

25. *The Case for Better Governance of Children's Data: A Manifesto.* UNICEF, 2021, https://www.unicef.org/innocenti/media/1031/file/ UNICEF%20Global%20Insight%20Data%20Governance%20Mani festo.pdf.

5. Why We Should Think of Our Data as Currency

1. Turow, Joseph, et al. *Americans Can't Consent to Companies' Use of Their Data: They Admit They Don't Understand It, Say They're Helpless to Control It, and Believe They're Harmed When Firms Use Their Data— Making What Companies Do Illegitimate.* 15 Feb. 2023, https://papers. ssrn.com/sol3/papers.cfm?abstract_id=4391134.

2. Kelly, Girard, et al. *2023 State of Kids' Privacy: Who Is Monetizing Our Data?* Common Sense Media, 2023, https://www.commonsensemedia. org/sites/default/files/research/report/common-sense-media-2023-state-of-kids-privacy_0.pdf.

3. Zuboff, Shoshana. *The Age of Surveillance Capitalism: The Fight for a Human Future at the New Frontier of Power.* 2019, https:// shoshanazuboff.com/book/.

4. "Victory at Last! NY Attorney General Enforces Law and Makes College Board Stop Selling Student Data!" *Student Privacy Matters*, 19 Feb. 2024, https://studentprivacymatters.org/victory-at-last-ny-attorney-general-enforces-law-and-makes-college-board-stop-selling-student-data/.

5. "Attorney General James and NYSED Commissioner Rosa Secure $750,000 from College Board." New York State Office of the Attorney General, 19 Mar. 2024, https://ag.ny.gov/press-release/2024/attorney-general-james-and-nysed-commissioner-rosa-secure-750000-college-board.

6. *Data Broker Market: Global Industry Analysis and Forecast* (2025–2032) *by Data Category, Data Type, End-User, and Region.* Maximize Market Research (MMR), 2024, https://www.maximizemarketresearch.com/market-report/global-data-broker-market/55670/.

7. "FTC Takes Action Against General Motors for Sharing Drivers' Precise Location and Driving Behavior Data." Federal Trade Commission, 16 Jan. 2025, https://www.ftc.gov/news-events/news/press-releases/2025/01/ftc-takes-action-against-general-motors-sharing-drivers-precise-location-driving-behavior-data.

8. Dachwitz, Ingo, and Sebastian Meineck. "New data set reveals 40,000 apps behind location tracking." *netzpolitik.org*, 15 Jan. 2025, https://netzpolitik.org/2025/databroker-files-new-data-set-reveals-40000-apps-behind-location-tracking/.

9. *Data Brokers, A Call for Transparency and Accountability.* Federal Trade Commission, May 2014, https://www.ftc.gov/system/files/documents/reports/data-brokers-call-transparency-accountability-report-federal-trade-commission-may-2014/140527databrokerreport.pdf.

10. *AI Training Dataset Market Size, Share, and Industry Analysis.* Fortune Business Insights, 2024, https://www.fortunebusinessinsights.com/ai-training-dataset-market-109241.

11. Tsukayama, Hayley. "Why Getting Paid for Your Data Is a Bad Deal." Electronic Frontier Foundation, 26 Oct. 2020, https://www.eff.org/deeplinks/2020/10/why-getting-paid-your-data-bad-deal.

12. *FTC Report Shows Rise in Sophisticated Dark Patterns Designed to Trick and Trap Consumers.* Federal Trade Commission, 15 Sept. 2022, https://www.ftc.gov/news-events/news/press-releases/2022/09/ftc-report-shows-rise-sophisticated-dark-patterns-designed-trick-trap-consumers.

13. "Tech Megacaps Plan to Spend More Than $300 Billion in 2025 as AI Race Intensifies." CNBC, 8 Feb. 2025, https://www.cnbc.com/2025/02/08/tech-megacaps-to-spend-more-than-300-billion-in-2025-to-win-in-ai.html.

14. *Federal Lobbying.* Open Secrets, 2024, https://www.opensecrets.org/federal-lobbying/top-spenders?cycle=2024.

15. Vong, Wai Keen, et al. "Grounded Language Acquisition through the

Eyes and Ears of a Single Child." *Science*, 1 Feb. 2024, https://www.science.org/doi/10.1126/science.adi1374.

16. *The Case for Better Governance of Children's Data: A Manifesto.* UNICEF, 2021, https://www.unicef.org/innocenti/media/1031/file/UNICEF%20Global%20Insight%20Data%20Governance%20Manifesto.pdf.

17. LeVasseur, Lisa. *2022 K-12 EdTech Safety Benchmark Findings Report 3.* Internet Safety Labs, 2024, https://internetsafetylabs.org/wp-content/uploads/2024/02/2022-K-12-EdTech-Safety-Benchmark-Findings-Report-3.pdf.

18. Allyn, Bobby. "TikTok to Pay $92 Million to Settle Class-Action Suit over 'Theft' of Personal Data." NPR, 25 Feb. 2021, https://www.npr.org/2021/02/25/971460327/tiktok-to-pay-92-million-to-settle-class-action-suit-over-theft-of-personal-data.

19. Figliola, Patricia Moloney. "TikTok: Technology Overview and Issues." *Congressional Research Service*, 30 June 2023, https://www.congress.gov/crs-product/R46543.

20. Page, Scott E. "Making the Difference: Applying a Logic of Diversity." *Academy of Management Perspectives*, Nov. 2007, http://www.jstor.org/stable/27747407.

21. Calvert, Luanne. *The Era of Mutts in a World of Pedigree.* TEDxBerlin, 2024, https://www.youtube.com/watch?v=W3US7PdlTy8.

22. *Diversity in AI.* Stanford University Human-Centered Artificial Intelligence, https://hai.stanford.edu/assets/files/2021-ai-index-report-_chapter-6.pdf.

23. Chilazi, Siri. *Advancing Gender Equality in Venture Capital: What the Evidence Says About the Current State of the Industry and How to Promote More Gender Diversity, Equality, and Inclusion.* Harvard Kennedy School, https://www.hks.harvard.edu/centers/wappp/publications/advancing-gender-equality-venture-capital.

24. *Funding Black-Founded Startups: How VC Funds Are Missing Black Talent.* Columbia Business School, 10 July 2024, https://business.columbia.edu/research-brief/funding-black-founded-startups.

25. Weber, Dylan, et al. *Bounded Confidence: How AI Could Exacerbate Social Media's Homophily Problem.* 19 Oct. 2022, https://scholarworks.umb.edu/cgi/viewcontent.cgi?article=1827&context=nejpp.

26. Milmo, Dan. "'We Definitely Messed Up': Why Did Google AI Tool Make Offensive Historical Images?" *The Guardian*, 8 Mar. 2024, https://www.theguardian.com/technology/2024/mar/08/we-definitely-messed-up-why-did-google-ai-tool-make-offensive-historical-images.

27. Sukharevsky, Alexander, et al. *Seizing the Agentic AI Advantage.* McKinsey & Company, 13 June 2025, https://www.mckinsey.com/capabilities/quantumblack/our-insights/seizing-the-agentic-ai-advantage.

28. Zuboff, Shoshana. *The Age of Surveillance Capitalism: The Fight for a Human Future at the New Frontier of Power.* 2019, https://shoshanazuboff.com/book/.

6. How AI Can Challenge Kids' Mental Health

1. De Freitas, Julian, et al. *AI Companions Reduce Loneliness.* Harvard Business School, 2024, https://www.hbs.edu/ris/Publication%20Files/24-078_a3d2e2c7-eca1-4767-8543-122e818bf2e5.pdf.
2. "Will I Be Believed?" How Deepfakes Risk Eroding Kids' Confidence in the People around Them." *Thorn,* 26 Sept. 2024, https://www.thorn.org/blog.
3. Radivojevic, Kristina, et al. "Social Media Bot Policies: Evaluating Passive and Active Enforcement." *arXiv,* 27 Sept. 2024, https://arxiv.org/pdf/2409.18931.
4. Kurian, N. "'No, Alexa, No!': Designing Child-Safe AI and Protecting Children from the Risks of the 'Empathy Gap' in Large Language Models." *Learning, Media and Technology,* 2024, pp. 1–14, https://doi.org/10.1080/17439884.2024.2367052.
5. Yuan, Z., X. Cheng, and Y. Duan. "Impact of Media Dependence: How Emotional Interactions between Users and Chat Robots Affect Human Socialization." *Frontiers in Psychology,* vol. 15, 16 Aug. 2024, https://doi.org/10.3389/fpsyg.2024.1388860.
6. Nguyen, Stephanie, and Erie Meyer. "Tech Brief: AI Sycophancy & OpenAI." Georgetown Law Technology Institute, https://www.law.georgetown.edu/tech-institute/insights/tech-brief-ai-sycophancy-openai-2/.
7. Nguyen, C. Thi. "Echo Chambers and Epistemic Bubbles." *Episteme,* vol. 17, no. 2, 2020, pp. 141–161, https://doi.org/10.1017/epi.2018.32.
8. "Taylor Swift deepfakes spread online, sparking outrage." CBS News, 26 Jan. 2024, https://www.cbsnews.com/news/taylor-swift-deepfakes-online-outrage-artificial-intelligence/.
9. Mani, Francesca. "Anti-deepfake activist." *TIME100 AI 2024,* https://time.com/7012803/francesca-mani/.
10. *Children and Deepfakes.* European Parliament, 2025, https://www.europarl.europa.eu/RegData/etudes/BRIE/2025/775855/EPRS_BRI(2025)775855_EN.pdf.
11. *Generative Adversarial Networks (GANs).* IBM, https://www.ibm.com/think/topics/generative-adversarial-networks.
12. O'Brien, Matt, and Ali Swenson. "Tech Giants Sign Voluntary Accord to Combat Election Deepfakes Generated with AI." PBS NewsHour, 16 Feb. 2024, https://www.pbs.org/newshour/nation/tech-giants-sign-voluntary-accord-to-combat-election-deepfakes-generated-with-ai.

13. DeStefano, Jennifer. *Victim of AI Deepfake Kidnapping/Extortion Scam.* U.S. Senate Judiciary, Subcommittee on Human Rights and the Law, 13 June 2023, https://www.judiciary.senate.gov/committee-activity/hearings/artificial-intelligence-and-human-rights.

14. "Beware of Virtual Kidnapping Ransom Scam." National Institute of Health, https://ors.od.nih.gov/News/Pages/Beware-of-Virtual-Kidnapping-Ransom-Scam.aspx.

15. Wei, Marlynn. "The Psychological Effects of AI Clones and Deepfakes." *Psychology Today*, 13 Feb. 2024, https://www.psychologytoday.com/us/blog/urban-survival/202401/the-psychological-effects-of-ai-clones-and-deepfakes.

16. "Our Epidemic of Loneliness and Isolation." Office of the U.S. Surgeon General, 2023, https://www.hhs.gov/sites/default/files/surgeon-general-social-connection-advisory.pdf.

17. *The eSafety Guide.* eSafety Commissioner, Australian Government, https://www.esafety.gov.au/key-topics/esafety-guide.

18. *AI Studio.* Meta, https://ai.meta.com/ai-studio/.

19. Holyoak, Melissa. *Statement of Commissioner Melissa Holyoak Regarding Issuance of Section 6(b) Orders to Companies That Offer Generative AI Companion Products or Services.* Federal Trade Commission, 11 Sept. 2025, https://www.ftc.gov/system/files/ftc_gov/pdf/statement-holyoak-ai-6b-study_0.pdf.

20. Horwitz, Jeff. "Meta's AI Rules Have Let Bots Hold 'Sensual' Chats with Kids, Offer False Medical Info." *The Wall Street Journal*, 14 Aug. 2025, https://www.reuters.com/investigates/special-report/meta-ai-chatbot-guidelines/.

21. *Talk, Trust, and Trade-Offs: How and Why Teens Use AI Companions.* Common Sense Media, 2025, https://www.commonsensemedia.org/sites/default/files/research/report/talk-trust-and-trade-offs_2025_web.pdf.

22. Ryder, Bethanie. "AI Friends, 'Gu Liao,' Paperclip Love: What China's Gen Z Wants." *Jing Daily*, 12 Jan. 2025, https://jingdaily.com/posts/ai-friends-gu-liao-paperclip-love-what-china-gen-z-want.

23. Mahari, Robert, and Pat Pataranutaporn. "Addictive Intelligence: Understanding Psychological, Legal, and Technical Dimensions of AI Companionship." *MIT Case Studies in Social and Ethical Responsibilities of Computing*, Winter 2025, https://doi.org/10.21428/2c646de5.2877155b.

24. Darvishi, Ali, et al. "Impact of AI Assistance on Student Agency." *Computers & Education*, vol. 210, 2024, 104967, https://doi.org/10.1016/j.compedu.2023.104967.

25. Zhao, Jun. "Designing for Children's Autonomy in the Age of AI (Part I)." Oxford Child-Centred AI (OxfordCCAI) Design Lab, 16 May 2024, https://oxfordccai.org/blog/20-24-5-agency/.

26. Rao, G. P., et al. "Developing Resilience and Harnessing Emotional Intelligence." *Indian Journal of Psychiatry*, vol. 66, suppl. 2, Jan. 2024, https://doi.org/10.4103/indianjpsychiatry.indianjpsychiatry_601_23.

27. "About." *Right Mindset PLLC*, https://www.rightmindsetpllc.com/about.

28. Dubber, Markus D., Frank Pasquale, and Sunit Das, editors. *The Oxford Handbook of Ethics of AI*. Oxford University Press, 2020.

7. How to Have a Say in Shaping the Future of AI

1. "The Clinton/Gore Administration: Investing in America's Future — 21st Century Education." The White House, 23 Nov. 1998, https://clintonwhitehouse4.archives.gov/WH/Work/112398.html.

2. "Advancing Artificial Intelligence Education for American Youth." The White House, 23 Apr. 2025, https://www.whitehouse.gov/presidential-actions/2025/04/advancing-artificial-intelligence-education-for-american-youth/.

3. Gelernter, David. "Should Schools Be Wired to the Internet?" *Time*, 25 May 1998, https://time.com/archive/6732843/dave-gelernter-should-schools-be-wired-to-the-internet/.

4. *The EdTech Top 40: K-12 EdTech Engagement During the 2024-25 School Year*. Instructure, 2025, https://www.instructure.com/resources/research-reports/edtech-top40-2025.

5. "Chromebook Expiration Dates and How They Are Affecting School Districts." *BuyBoard*, https://www.buyboard.com/insights/chromebook-expiration-dates-and-how-they-are-affecting-school-districts.

6. Holtermann, Callie. "A New Headache for Honest Students: Proving They Didn't Use A.I." *The New York Times*, 17 May 2025, https://www.nytimes.com/2025/05/17/style/ai-chatgpt-turnitin-students-cheating.html.

7. *AI Detector Market*. MarketsandMarkets, https://www.marketsandmarkets.com/Market-Reports/ai-detector-market-199981626.htm.

8. "Impact of AI on Student Assessment and Academic Integrity." *Education Sciences*, vol. 14, no. 11, 2024, https://www.mdpi.com/2227-7102/14/11/1197.

9. "Exploring Children's Rights and A.I." *Children's Parliament Scotland*, 2023, https://www.childrensparliament.org.uk/our-work/exploring-childrens-rights-and-ai/.

10. Spilka, Dmytro. "How Scotland's New Supercomputer Project Will Prepare UK Sectors for New AI Frontiers." *The Scotsman*, 18 June 2025, https://www.scotsman.com/community/how-scotlands-new-supercomputer-project-will-prepare-uk-sectors-for-new-ai-frontiers-5183332.

11. "Enhancing Access and Inclusion through AI-Driven Pedagogies." Chartered Association of Business Schools, https://charteredabs.org/insights/

impact-case-studies/enhancing_access_and_inclusion_through_ai-driven_pedagogies.

12. Fung, Archon. "Varieties of Participation in Complex Governance." *Public Administration Review, vol. 66, Special Issue: Collaborative Public Management*, Dec. 2006, http://www.jstor.org/stable/4096571.

13. *Cities and Artificial Intelligence Task Forces*. National League of Cities, https://www.nlc.org/resource/cities-and-artificial-intelligence-task-forces.

14. "Navigating Technological Challenges." Pew Research Center, 1 Sept. 2021, https://www.pewresearch.org/internet/2021/09/01/navigating-technological-challenges/.

15. "Guidance on the Use of Federal Grant Funds to Improve Education Outcomes Using Artificial Intelligence." U.S. Department of Education, 22 July 2025, https://www.ed.gov/media/document/opepd-ai-dear-colleague-letter-7222025-110427.pdf.

16. *Empowering Communities: Public Libraries, Inclusive Civic Engagement, and Artificial Intelligence*. Center for Technology in Government, University at Albany, 2024, https://www.ctg.albany.edu/media/pubs/pdfs/IMLS_Best_Practices_Report.pdf.

17. *2025 Parent Power Agenda*. National Parents Union, 16 Aug. 2024, https://nationalparentsunion.org/2024/08/16/2025-parent-power-agenda/.

18. "About Us." Student Privacy Matters, https://studentprivacymatters.org/about-us/.

19. *Mozilla Foundation. Responsible Computing Challenge and MozFest*. Mozilla Foundation, 2024, https://foundation.mozilla.org.

20. "NEA delegates approve measure to recommend policy actions on use of AI." National Education Association, https://www.nea.org/about-nea/media-center/press-releases/nea-delegates-approve-measure-recommend-policy-actions-use-ai.

8. What Excellent AI Support at Home Can Look Like

1. Goleman, Daniel. *Emotional Intelligence: Why It Can Matter More Than IQ*. Bantam Books, 1995.

2. Orta, I. M., and S. M. Camgoz. "Exploring emotional intelligence at work: A review of current evidence." *Social issues in the workplace: Breakthroughs in research and practice, IGI Publishing/IGI Global*, 2018, https://doi.org/10.4018/978-1-5225-3917-9.ch009.

3. Search Inside Yourself Leadership Institute, https://siyli.org/about.

4. *Future of Jobs Report 2025*. World Economic Forum, Jan. 2025, https://reports.weforum.org/docs/WEF_Future_of_Jobs_Report_2025.pdf.

5. Gerlich, Michael. "AI Tools in Society: Impacts on Cognitive Offloading

and the Future of Critical Thinking." *Societies*, vol. 15, no. 1, 3 Jan. 2025, 6, https://doi.org/10.3390/soc15010006.

6. Marano, G., et al. "The Neuroscience Behind Writing: Handwriting vs. Typing-Who Wins the Battle?" *Life*, vol. 15, no. 3, 22 Feb. 2025, p. 345, https://doi.org/10.3390/life15030345.

7. "Tech Ethics Center Initiatives." Stanford Ethics in Society, https://ethicsinsociety.stanford.edu/tech-ethics/tech-ethics-center-initiatives.

8. "Philosophy, Politics and Economics." University of Oxford, 2025, https://www.ox.ac.uk/admissions/undergraduate/courses/course-listing/philosophy-politics-and-economics.

9. "Interdisciplinary Science: The Next Big Breakthrough?" World Economic Forum, 13 Nov. 2023, https://www.weforum.org/stories/2023/11/interdisciplinary-science-next-big-breakthrough/.

9. How We Can Build a Better Future Together

1. *Humanoids: A $5 Trillion Market*. Morgan Stanley, 14 May 2025, https://www.morganstanley.com/insights/articles/humanoid-robot-market-5-trillion-by-2050.

2. *Humanoid Robot Market*. Future Market Insights, 2025, https://www.futuremarketinsights.com/reports/humanoid-robot-market. Accessed 18 Sept. 2025.

3. Koetsier, John. "Humanoid Robots: Friends, Not Just Workers." *Forbes*, 20 Jan. 2025, https://www.forbes.com/sites/johnkoetsier/2025/01/20/humanoid-robots-friends-not-just-workers/.

4. "New Survey Uncovers Most Americans Spend More Than Ten Hours Every Day Online." *Altice USA, Optimum*, 27 July 2025, https://www.alticeusa.com/news/articles/new-survey-uncovers-most-americans-spend-more-ten-hours-every-day-online.

5. "The Real Cost of Doomscrolling." Payless Power, 2025, https://paylesspower.com/blog/the-real-cost-of-doomscrolling/.

6. Haidt, Jonathan. *The Anxious Generation: How the Great Rewiring of Childhood Is Causing an Epidemic of Mental Illness*. 2024, https://www.anxiousgeneration.com/book.

7. Wikipedia contributors. "Do You Know Where Your Children Are?" Wikipedia, https://en.wikipedia.org/wiki/Do_you_know_where_your_children_are%3F.

8. "Why I Let My 9-Year-Old Ride Subway Alone." *The New York Sun*, 6 May 2022, https://web.archive.org/web/20220506154236/https://www.nysun.com/article/opinion-why-i-let-my-9-year-old-ride-subway-alone.

9. The Alpha School, https://alpha.school.

10. Paul, Annie Murphy. *The Extended Mind: The Power of Thinking*

Outside the Brain. 2021, https://anniemurphypaul.com/books/the-extended-mind/.

11. "AI Will Transform Teaching and Learning: Let's Get It Right." Stanford HAI, 9 Mar. 2023, https://hai.stanford.edu/news/ai-will-transform-teaching-and-learning-lets-get-it-right.

12. Eastwood, Brian. "When Humans and AI Work Best Together — and When Each Is Better Alone." MIT Sloan Ideas Made to Matter, 3 Feb. 2025, https://mitsloan.mit.edu/ideas-made-to-matter/when-humans-and-ai-work-best-together-and-when-each-better-alone.

13. Jibril, Halima. "ChatGPT Might Be Making Us Lazy and Dumb." *Dazed Digital*, 19 June 2025, https://www.dazeddigital.com/life-culture/article/68060/1/chatgpt-might-be-making-lazy-and-dumb-generative-ai-chatbot.

14. Kosmyna, Nataliya, et al. "Your Brain on ChatGPT: Accumulation of Cognitive Debt When Using an AI Assistant for Essay Writing Task." *arXiv preprint*, 2025, https://arxiv.org/abs/2506.08872.

15. Kosmyna, Nataliya. LinkedIn post, 2023, https://www.linkedin.com/posts/nataliekosmina_mit-ai-brain-activity-7340386826504876033-X45W.

16. Banks, Sarah A. *A Historical Analysis of Attitudes Toward the Use of Calculators in Junior High and High School Math Classrooms in the United States Since 1975.* Master's thesis, Cedarville University, 2008, https://files.eric.ed.gov/fulltext/ED525547.pdf.

17. Nickl, Peter, et al. "The Evolution of Online News Headlines." *Humanities and Social Sciences Communications*, Mar. 2025, https://doi.org/10.1057/s41599-025-04514-7.

18. "Local News Landscape." Northwestern University Medill School of Journalism, 2025, https://localnewsinitiative.northwestern.edu/projects/state-of-local-news/explore/#/localnewslandscape.

19. "Poll Finds Bipartisan Agreement on a Key Issue: Regulating AI." *MinnPost*, 17 July 2025, https://www.minnpost.com/community-voices/2025/07/ai/.

20. Cox, Daniel A. *The State of American Friendship: Change, Challenges, and Loss.* American Survey Center, 8 June 2021, https://www.americansurveycenter.org/research/the-state-of-american-friendship-change-challenges-and-loss/.

21. "Social media use increases depression and loneliness." *Penn Today*, https://penntoday.upenn.edu/news/social-media-use-increases-depression-and-loneliness.

22. Sidoti, Olivia, and Emily A. Vogels. *What Americans Know About AI, Cybersecurity and Big Tech.* Pew Research Center, 17 Aug. 2023, https://www.pewresearch.org/internet/2023/08/17/what-americans-know-about-ai-cybersecurity-and-big-tech/.

23. "The Nobel Prize in Chemistry 2024." NobelPrize.org, https://www.nobelprize.org/prizes/chemistry/2024/press-release/.

24. *AlphaFold*. Google DeepMind, https://deepmind.google/science/alphafold/.
25. *The Thinking Game*. Roco Films, https://rocofilms.com/films/the-thinking-game/.
26. *AI Tools*. National University, https://resources.nu.edu/AI-at-NU/AItools.
27. "Efficiently Matching Patients with Organs for Transplant." United Network for Organ Sharing (UNOS), 2023, https://unos.org/news/efficiently-matching-patients-with-organs-for-transplant/.
28. "Biodiversity and Conservation." *Biological Conservation*, 2023, https://www.sciencedirect.com/science/article/pii/S0006320723001921.
29. "Nature and AI." Nature Scientific Reports, 2025, https://www.nature.com/articles/s41598-025-16014-4.
30. "MIT Solve Solutions." MIT Solve, https://solve.mit.edu/solutions/79429.